Moon Landing

Moon Landings

Bases & Exploration

A History & Current Plans

By Martin K. Ettington

Table of Contents

1.0 Introduction 10

2.0 Earth and Lunar Environments 12

2.1 The Moon Environment...12

2.2 The Discovery of Ice at the Lunar Poles 14

3.0 Legal Issues and Treaties on the Moon..................................19

4.0 Requirements for Building Moon Bases.................................24

5.0 Moon Bases in Science Fiction 27

5.1 From the Earth to the Moon by Jules Verne 27

5.2 The First Man in the Moon by H.G. Wells.............................30

5.3 The Moon is a Harsh Mistress By Robert Heinlein 32

5.4 The 2001 Movie Moon Base by Arthur C. Clarke..................34

5.5 Artemis By Andy Weir 36

6.0 Early Moon Base Design Proposals 38

6.1 Project Horizon..40

6.2 Subsurface Moon bases..42

6.3 The LESA Moon Base 44

6.4 Underground Base Plan..47

7.0 The First Moon Landings 49

8.0 NASA's Plans for Lunar Travel, Bases & More 52

8.2 The Deep Space Gateway...66

8.3 Modern Lunar Lander Concepts...70

8.4 SpaceX's Starship Lunar Lander 75

8.5 Settling the South Pole...78

8.6 Far Side Bases 79

9.0 Types of Moon Bases 80

9.2 Printing Buildings..84

9.3 Underground Structures 86

10.0 Lunar Space Suits 88

11.0 Lunar Transportation 99

12.0 Lunar Power Production 104

13.0 Communications...108

14.0 Mining Water Ice 120
15.0 Growing Food on the Moon 124
16.0 Current Plans for NASA Moon Bases 137
17.0 Chinese and Russian Lunar Plans........................146
17.1 Unmanned Lunar Lander Visits to the Farside 147
17.2 A Chinese and Russian Lunar Base.......................149
18.0 Additional Technologies Needed........................152
19.0 Schedules for Construction 165
20.0 Atreus Countries, Companies, and Plans.............167
20.1 Space Launch System 168
20.2 The Orion Capsule..170
20.3 Lunar Gateway Partners 173
24.0 Moon Base Cost Estimates 176
25.0 A Story of Building the Moon Base 179
25.1 Early Moon Base Construction 179
25.2 Main Shelter Construction....................................183
25.3 Finishing the Shelter 188
26.0 Types of People Needed..191
27.0 Reasons for a Moon Base.......................................194
27.2 A Dry run for Visiting Mars...................................196
27.3 Exploitation of Lunar Resources..........................197
28.0 Moon Base Future Growth Plans..........................199
29.0 Future Moon Structures 202
30.0 Summary 204
31.0 Bibliography..205
32.0 Index 208
Other books by Martin K. Ettington 212

NASA's plans to land on the Moon again have been in flux in the last few years. Russia is also no longer a partner as they were previously.

What is the history of plans to build a base on the Moon and what are all the issues involved?

Where should the base be built and what will be do once we have long term habitation on the Moon?

And what are the latest plans for lunar rovers, communications, and other moon base systems?

My hope is that the reader will learn about all of the issues important to building a Moon Base to better understand what will be required.

You might even want to become a Moon Base resident!

1.0 Introduction

This is an updated book about Moon bases and living on the Moon. I like to write about popular topics which interest me and the Moon is one of them.

Maybe it has to do with my growing up in the 1960s at the height of the Apollo Program. I remember walking to school at about seven years old and the kids in the group were all talking about the Mercury Astronauts and how cool they were. Then I followed all of the succeeding Gemini and Apollo missions. It was an incredible experience to watch the Apollo 11 landing and first Moonwalk at Boy Scout camp with us all gathered in the dining hall.

Later on I worked for Hewlett Packard at the NASA Human Spaceflight Center in Houston for a couple of years in the mid nineteen eighties. I also met many of the Astronauts including July Resnick who was killed on the Challenger Shuttle when it blew up at launch. Watching President Reagans' speech at the Johnson Space Center as a memorial to them was also quite the experience. One of my friends who took me up in his Pitt Special acrobatic plane a couple of times later became an Astronaut on the MIR, Space Shuttle, and Space Station. I got my Pilots license at Laporte Airport, Texas where many of the off duty Astronauts flew acrobatic planes for fun.

I also applied to the Astronaut Corp but being a civilian and not having any degrees beyond my B.S. in Engineering Science filtered me out. Never the less, I've always been a fan of the space program. I still follow all the details I can to this day.

In the last couple of years I wrote a comprehensive guide to Space Colonies titled "Designing and Building Space Colonies-A Blueprint for the Future". In that book I covered a lot of the issues about living in space and construction of space colonies.

With the birth of the Atreus Program and plans to land man on the Moon again in 2025 or later, I thought this would be a good time to go into the history of Moon landings and proposed Moon colonies and maybe make a few of my own suggestions too.

I hope you enjoy this journey into our near future.

2.0 Earth and Lunar Environments

The Moon and the Earth are very different environments. It is important to emphasize the differences in the living conditions of these two spheres to better understand what will have to be built into structures to live safely on the Moon.

2.1 The Moon Environment

Gravity

The Moon's mass is about 1/80th (1.2%) of the Earth's mass, so the Moon's gravity is much less than the Earth's gravity; specifically, the Moon's gravity is 1/6th (16.7%) of the Earth's gravity. Or, stated another way, the Moon's gravity is 5/6 (83.3%) LESS than the Earth's.

This means that you will weigh much less on the Moon but the inertia of objects in motion will remain the same. So you can lift a lot, but will still have trouble stopping a large object in motion.

Atmosphere

On Earth at sea level the atmospheric pressure is 14.7 pounds per square inch. The Moon doesn't have an atmosphere and is basically a vacuum. This means all structures will need to be airtight and have airlocks for going in and out. Of course you will need to wear a spacesuit on the surface on the Moon.

Radiation

The surface of the Moon is baldly exposed to cosmic rays and solar flares, and some of that radiation is very hard to stop with shielding. Furthermore, when cosmic rays hit the ground, they produce a dangerous spray of secondary particles right at your feet. All this radiation penetrating human flesh can damage DNA, boosting the risk of cancer and other maladies. Spacesuits and buildings need to be able to protect humans from this radioactive environment.

Of course the atmosphere on Earth protects us from these deadly rays.

Temperature

Daytime on one side of the Moon lasts about 13 and a half days, followed by 13 and a half nights of darkness. When sunlight hits the Moon's surface, the temperature can reach 260 degrees Fahrenheit (127 degrees Celsius). When the sun goes down, temperatures can dip to minus 280 F (minus 173 C)

Earth's temperature varies from the extremes of 132 degree Fahrenheit to minus 126.6 degrees Fahrenheit.

These temperature extremes also mean that your spacesuit needs to be designed to withstand large temperature shifts.

Water

The Earth is a watery planet which is covered 71 percent by water in Oceans and lakes.

We used to think that there was no water on the Moon. Now we have satellites which show there should be water ice at the Poles, especially the South Pole. The question is can we mine it and purify it?

2.2 The Discovery of Ice at the Lunar Poles

Finding water ice on the moon is of great importance for building moon bases. This will provide oxygen and water for the base's consumption and also hydrogen which can be used a rocket fuel for many rockets.

Ice at the moon's poles might have come from ancient volcanoes. The eruptions may have produced several transient atmospheres

Schrödinger crater (shown) lies near the moon's South Pole. Ice might have arrived at both lunar poles as water vapor released by ancient volcanic eruptions.

Four billion years ago, lava spilled onto the moon's crust, etching the man in the moon we see today. But the volcanoes may have also left a much colder legacy: ice.

Two billion years of volcanic eruptions on the moon may have led to the creation of many short-lived atmospheres, which contained water vapor, a new study suggests. That vapor could have been transported through the atmosphere before settling as ice at the poles, researchers report in the May Planetary Science Journal.

Since the existence of lunar ice was confirmed in 2009, scientists have debated the possible origins of water on the moon, which include asteroids, comets or electrically charged atoms carried by the solar wind Or, possibly, the water originated on the moon itself, as vapor belched by the rash of volcanic eruptions from 4 billion to 2 billion years ago.

"It's a really interesting question how those volatiles [such as water] got there," says Andrew Wilcoski, a planetary scientist at the University of Colorado Boulder. "We still don't really have a good handle on how much are there and where exactly they are."

Wilcoski and his colleagues decided to start by tackling volcanism's viability as a lunar ice source. During the heyday of lunar volcanism, eruptions happened about once every 22,000 years. Assuming that H_2O constituted about a third of volcano-spit gasses — based on samples of ancient lunar magma — the researchers calculate that the eruptions released upward of 20 quadrillion kilograms of water vapor in total, or the volume of approximately 25 Lake Superiors.

Some of this vapor would have been lost to space, as sunlight broke down water molecules or the solar wind blew the molecules off the moon. But at the frigid poles, some could have stuck to the surface as ice.

For that to happen, though, the rate at which the water vapor condensed into ice would have needed to surpass the rate at which the vapor escaped the moon. The team used a computer simulation to calculate and compare these rates. The simulation accounted for factors such as surface temperature, gas pressure and the loss of some vapor to mere frost.

About 40 percent of the total erupted water vapor could have accumulated as ice, with most of that ice at the poles, the team found. Over billions of years, some of that ice would have converted back to vapor and escaped to space. The team's simulation predicts the amount and distribution of ice that remains. And it's no small amount: Deposits could reach hundreds of meters at their thickest point, with the South Pole being about twice as icy as the North Pole.

The results align with a long-standing assumption that ice dominates at the poles because it gets stuck in cold traps that are so cold that ice will stay frozen for billions of years.

Moon ice

These results from a computer simulation depict the potential present-day distribution and thickness of ice at the lunar poles following volcanic eruptions 4 billion to 2 billion years ago. The South Pole (left) retains more ice because it has more cold traps than the North Pole (right). The dotted lines depict longitude and latitude.

Potential ice deposits at the moon's poles

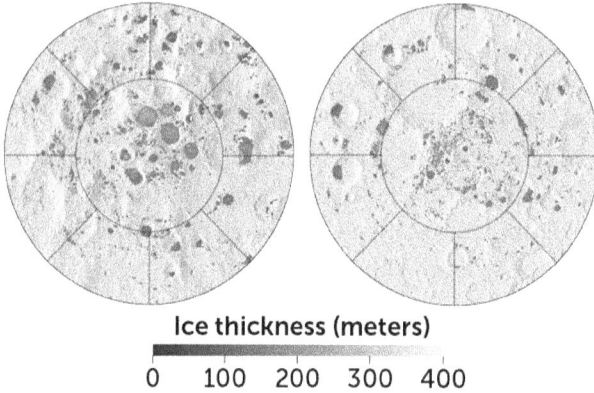

Ice thickness (meters)

0   100   200   300   400

A.X. WILCOSKI, P.O. HAYNE AND M.E.
LANDIS/PLANETARY SCIENCE JOURNAL 2022

"There are some places at the lunar poles that are as cold as Pluto," says planetary scientist Margaret Landis of the University of Colorado Boulder.

Volcanically sourced water vapor traveling to the poles, though, probably depends on the presence of an atmosphere, say Landis, Wilcoski and their colleague Paul Hayne, also a planetary scientist at the University of Colorado Boulder. An atmospheric transit system would have allowed water molecules to travel around the moon while also making it more difficult for them to flee into space. Each eruption triggered a new atmosphere, the new calculations indicate, which then lingered for about 2,500 years before disappearing until the next eruption some 20,000 years later.

This part of the story is most captivating to Parvathy Prem, a planetary scientist at Johns Hopkins Applied Physics Laboratory in Laurel, Md., who wasn't involved in the research. "It's a really interesting act of imagination.... How do you create atmospheres from scratch? And why do they sometimes go away?" she says. "The polar ices are one way to find out."

If lunar ice was belched out of volcanoes as water vapor, the ice may retain a memory of that long-ago time. Sulfur in the polar ice, for example, would indicate that it came from a volcano as opposed to, say, an asteroid. Future moon missions plan to drill for ice cores that could confirm the ice's origin.

Looking for sulfur will be important when thinking about lunar resources. These water reserves could someday be harvested by astronauts for water or rocket fuel, the researchers say. But if all the lunar water is contaminated with sulfur, Landis says, "That's a pretty critical thing to know if you plan on bringing a straw with you to the moon."

3.0 Legal Issues and Treaties on the Moon

Most legal issues regarding property usage on the Moon are derived from the United Nations Outer Space Treaty

<u>The United Nations and the Outer Space Treaty</u>

COPUOUS was established in 1958 and made permanent in 1959. As of mid-2016, it has 77 members, including major space-faring nations such as the United States (NASA), Russia (Roscosmos), Japan, China, Canada, Brazil, Australia and the member states of the European Space Agency,

The United Nations describes this committee as the "focal point" where international entities negotiate how to use space peacefully. COPUOUS' duties include exchanging information about space, keeping tabs on what government and nongovernmental organizations do in space, and promoting international cooperation. COPUOUS also formed two subcommittees in 1962 to deal with legal issues, and scientific and technical developments; secretariat services are provided by the United Nations Office for Outer Space Affairs (UNOOSA).

COPUOUS is the force behind five treaties and five principles that govern much of space exploration. The fundamental treaty is the Treaty on Principles Governing the Activities of States in the Exploration and Use of Outer Space, including the Moon and Other Celestial Bodies, or simply the "Outer Space Treaty." It was ratified in 1967, largely based on a set of legal principles the general assembly accepted in 1962.

The treaty has several major points to it. Some of the principal ones are:

Space is free for all nations to explore, and sovereign claims cannot be made. Space activities must be for the benefit of all nations and humans. (So, nobody owns the moon.)

Nuclear weapons and other weapons of mass destruction are not allowed in Earth orbit, on celestial bodies or in other outer-space locations. (In other words, peace is the only acceptable use of outer-space locations).

Individual nations (states) are responsible for any damage their space objects cause. Individual nations are also responsible for all governmental and nongovernmental activities conducted by their citizens. These states must also "avoid harmful contamination" due to space activities.

Treaties, principles and conferences

To support the Outer Space Treaty, four other treaties were put into place in the 1960s and 1970s to support peaceful space exploration. These treaties (referred to below by their nicknames) are:

The "Rescue Agreement" (1968), formed to give astronauts assistance during an unintended landing or when they are facing an emergency. States are told they "shall immediately take all possible steps to rescue them and render them all necessary assistance."

The "Liability Convention" (1972) outlines considerations if a space object causes damage or loss to human life. Its first article says, "A launching state shall be absolutely liable to pay compensation for damage caused by its space object on the surface of the earth or to aircraft flight."

The "Registration Convention" (1975), drawn up to help nations keep track of all objects launched into outer space. This United Nations registry is important for matters such as avoiding space debris. (For NASA, the United States Strategic Command gives real-time updates to the agency if space debris threatens a spacecraft or the International Space Station.)

The "Moon Agreement" (1979), which gives more detail on the Outer Space Treaty for property rights and usage of the moon and other celestial bodies in the solar system (except for objects that naturally enter the Earth from these bodies, namely, meteorites). This treaty, however, has only been signed by 16 nations, all of which are minor players in space exploration.

COPUOUS has also created five sets of principles to support these treaties.

The "Declaration of Legal Principles" (1963), from which the Outer Space Treaty was created in 1967, lays down guiding principles, including the idea that space exploration is for the benefit of all humans.

The "Broadcasting Principles" (1982) has to do with television broadcast signals. These principles include the idea of noninterference with other countries' signals, the provision of information to help with knowledge exchange, and the promotion of educational and social development (particularly in developing nations).

The "Remote Sensing Principles" (1986) concerns the use of electromagnetic waves to collect data on Earth's natural resources. Remote-sensing activities are supposed to be for all countries' benefit and should be carried out in the spirit of international cooperation.

The "Nuclear Power Sources Principles" (1992) concerns how to protect humans and other species from radiation if a launch goes awry, or a spacecraft flying by Earth accidently crashes to the surface. It's common for spacecraft exploring the outer solar system to use nuclear power sources for energy, since solar power is so weak out there.

The "Benefits Declaration" (1996) says that space exploration shall be carried out for the benefit of all states. This was created two years before the International Space Station — an effort of 15 nations — launched its first two modules into space.

The United Nations has also held three UNISPACE Conferences since 1968. (A fourth one will take place in 2018.) This is what each conference focused on or will focus on:

UNISPACE I (August 1968): Progress in space exploration, international cooperation and creating an "expert on space applications" within UNOOSA. The United Nations body then had several workshops in the 1970s on space applications such as remote sensing, telecommunications and cartography.

UNISPACE II/UNISPACE 82 (August 1982): Peaceful exploration of space (specifically, how to avoid an arms race). Following the conference, UNOOSA worked more closely with developing countries to develop their space technology capabilities.

UNISPACE III (July 1999): Protecting the space environment, giving developing countries more access to space and protecting Earth's environment. This led to the Vienna Declaration on Space and Human Development, with 33 recommendations for space-faring countries to follow. A follow-up report to the declaration was issued in 2004, five years after the conference.

UNISPACE+50 (2018): Will celebrate the 50th anniversary of the first UNISPACE conference and focus on what COPUOUS should do now that more nations and nongovernmental entities are exploring space.

4.0 Requirements for Building Moon Bases

Radiation Protection

The walls of any base structure will need to protect the inhabitants from normal ionizing radiation or even from much larger occasional solar flare radiation. This is a good reason to build the base underground or build a structure then bury it.

Power Sources

Power can be nuclear, solar, or some type of generator running hydrogen from water split into hydrogen and helium. One of these sources might be a fuel cell. There might need to be battery backup too. If the base is built in a South Pole crater like Shakelton crater then solar will be a problem since there is no Sun in the crater. One possibility is putting solar panels on the crater rim in the sunlight and running cables back to the base.

Atmosphere

Any shelter will need to provide an atmosphere where the residents can work in a shirtsleeve environment. There will also need to be safety protections in case of a puncture of the structure(s). This probably means having airlocks and/or building the shelter underground or covered by soil to reduce the possibility of loss of air.

Being able to get oxygen from cracking water into hydrogen and oxygen will help provide availability of this critical gas to breathe.

Water

Water will need to be provided and recycled to clean it. The ability to mine water ice, purify it, and melt it will be a great boon to the residents. It would be very costly to ship all the water needed from Earth's gravity well.

Temperature Control

The surface temperature varies from 260 degrees plus Fahrenheit to 280 minus degrees Fahrenheit needs to be insulated against in both shelters and spacesuits. This has already been addressed in Apollo spacesuit design and buildings would need to be built to these standards.

Vehicles for the surface

Vehicles moving around the Moon need to have a much longer range than during the Apollo missions and maybe also an enclosed cabin so travelers can get out of their suits and rest in a shirtsleeve environment.

Communications

If the lunar base is located at the South Pole in Shakelton Crater then it will not be in line of site with the Earth. Also, it will need to communicate with the Lunar Gateway circling the Moon. This probably necessitates having satellites in orbit around the Moon to connect everyone together.

Growing Food

It costs a lot of money to ship supplies to the Moon from Earth for a crew to stay on the Moon for extended periods. It would be much more cost effective to allow Moon workers to grow at least some of their own food hydroponically. This would also provide some fresh fun for the persons growing the food.

Sewage

Having some way to recycle or process sewage would also make sense. It might even be useful to help grow some crops.

5.0 Moon Bases in Science Fiction

There are many science fiction novels about the Moon but I've picked these five which in many ways have affected our thinking and plans for visiting the Moon.

5.1 From the Earth to the Moon by Jules Verne

The first modern story about visiting the Moon was Jules Verne's book "From the Earth to the Moon" written in 1865. It was followed by his sequel "Around the Moon"

Jules Verne was very prescient about several things such as:

A Moon Launch from Florida

Three men in the Capsule

About weightlessness and the vacuum of space.

This story seems to have led to the modern interest in travel to the Moon.

Here is a summary of the second book "Around the Moon"

Having been fired out of the giant Columbiad space gun, the Baltimore Gun Club's bullet-shaped projectile, along with its three passengers, Barbicane, Nicholl and Michael Ardan, begins the five-day trip to the Moon. A few minutes into the journey, a small, bright asteroid passes within a few hundred yards of them, but does not collide with the projectile. The asteroid had been captured by the Earth's gravity and had become a second moon.

The three travelers undergo a series of adventures and misadventures during the rest of the journey, including disposing of the body of a dog out a window, suffering intoxication by gases, and making calculations leading them, briefly, to believe that they are to fall back to Earth. During the latter part of the voyage, it becomes apparent that the gravitational force of their earlier encounter with the asteroid has caused the projectile to deviate from its course.

The projectile enters lunar orbit, rather than landing on the Moon as originally planned. Barbicane, Ardan and Nicholl begin geographical observations with opera glasses. The projectile then dips over the northern hemisphere of the Moon, into the darkness of its shadow. It is plunged into extreme cold, before emerging into the light and heat again. They then begin to approach the Moon's southern hemisphere. From the safety of their projectile, they gain spectacular views of Tycho, one of the greatest of all craters on the Moon. The three men discuss the possibility of life on the Moon, and conclude that it is barren. The projectile begins to move away from the Moon, towards the 'dead point' (the place at which the gravitational attraction of the Moon and Earth becomes equal). Michel Ardan hits upon the idea of using the rockets fixed to the bottom of the projectile (which they were originally going to use to deaden the shock of landing) to propel the projectile towards the Moon and hopefully cause it to fall onto it, thereby achieving their mission.

When the projectile reaches the point of neutral attraction, the rockets are fired, but it is too late. The projectile begins a fall onto the Earth from a distance of 260,000 kilometers (160,000 mi), and it is to strike the Earth at a speed of 185,400 km/h (115,200 mph), the same speed at which it left the mouth of the Columbiad. All hope seems lost for Barbicane, Nicholl and Ardan. Four days later, the crew of a US Navy vessel, Susquehanna, spots a bright meteor fall from the sky into the sea. This turns out to be the returning projectile, and the three.

5.2 The First Man in the Moon by H.G. Wells

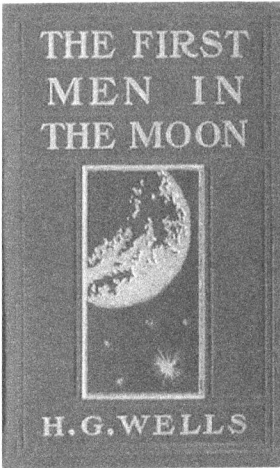

Originally published in 1901, the novel has been re-released 100 years later. The publishers understand that presentation is just as important as the overall story and the book greets the reader with a stunning cover illustration by Chris Moore, an artist trained at the Royal College of Art, who proves more than capable of capturing the essence of this adventure.

Set in England at the beginning of the 20th century, average industrialist Bedford finds himself entwined in the machinations of Cavor, an eccentric genius who has developed Cavorite, a substance that negates the pull of gravity. The two men construct a vessel called the Sphere which hurls them to the Moon. But the adventurers have very different agendas. Cavor hopes to discover a utopian society he imagines living on the planet, while Bedford is purely interested in the monetary gain the trip represents (after all, everyone knows there's gold on the Moon). Once they arrive, they stumble upon the world of the Selenites, insect-like, biologically engineered aliens living beneath the surface of the Moon in dark, cavernous, technologically-astounding cities. Then things go drastically wrong...

Wells' envisioning of Earth's satellite is fascinating in its accuracy; a barren planet with a thin (yet breathable) atmosphere, a freezing night and very little gravity. However, when the sun rose, Wells imagined forests of trees and plants exploding to life, having a mere Moon-day (which is like an Earth week) to grow, germinate and seed before the cold of the night withers them. Wells saw the possibility that the Moon itself would be full of catacombs, tunnels and internal seas. The Selenite society (although Cavor humorously refers to them as "Moonies") would exist beneath the surface like an ant colony. The images he creates are briefly seen by Bedford and somewhat described later by Cavor, which Wells has cleverly done to leave the reader's imagination to paint its own picture of this underworld.

5.3 The Moon is a Harsh Mistress By Robert Heinlein

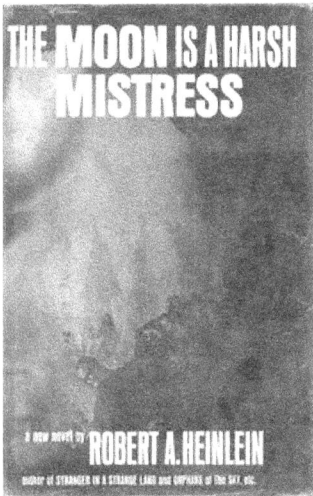

This is my favorite Science Fiction Novel of all time. At the time of the story, 2075, the Moon (Luna) is used as a penal colony by Earth's government, with the inhabitants living in underground cities. Most inhabitants (called "Loonies") are criminals, political exiles, or their descendants. The total population is about three million, with men outnumbering women two to one, so that polyandry is the norm. Although Earth's Protector of the Lunar Colonies (called the "Warden") holds power, in practice, little intervention exists in the loose lunar society.

Due to the low surface gravity of the Moon, Loonies who stay longer than a few months undergo "irreversible physiological changes and can never again live in comfort and health in a gravitational field six times greater than that to which their bodies have become adjusted".

HOLMES IV ("High-Optional, Logical, Multi-Evaluating Supervisor, Mark IV") is the Lunar Authority's master computer, having almost total control of Luna's machinery on the grounds that a single computer is cheaper than (though not as safe as) multiple independent systems.

The story is narrated by Manuel Garcia "Mannie" O'Kelly-Davis, a computer technician who discovers that HOLMES IV has achieved self-awareness and has developed a sense of humor. Mannie names it "Mike" after Mycroft Holmes, brother of Sherlock Holmes, and they become friends.

The main story is about how moon residents conduct a successful revolution to get their independence from Earth.

This story also includes a linear accelerator used to launch payloads into orbit and even to bombard the Earth with rocks.

5.4 The 2001 Movie Moon Base by Arthur C. Clarke

(Movie by Stanley Kubrick)

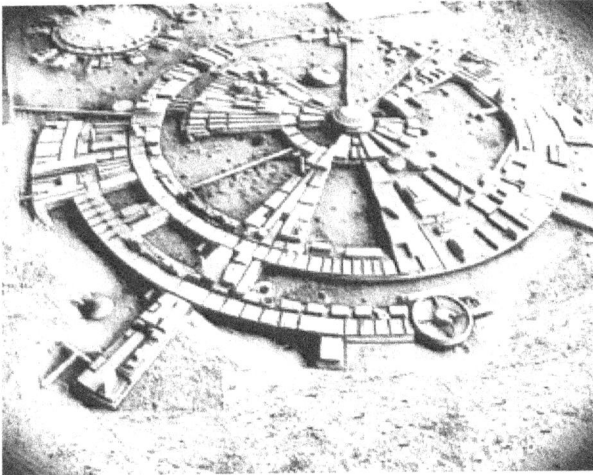

2001 was a movie based on the book 2001 by Arthur C. Clarke. In this book a monolith is found on the Moon emitting a strong signal which another spaceship eventually traces to Jupiter and finds a super race of aliens.

As part of the movie there is vivid trip to a base underground on the Moon which is also shown in this reddish picture below.

This picture is part of a sequence where the Moon landing vehicle is descending on an elevator into the underground base.

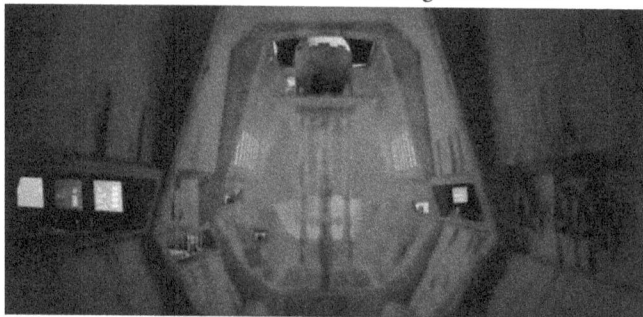

It is a very realistic portrayal of a base underground on the Moon.

5.5 Artemis By Andy Weir

Andy Weir is the Author of an incredible story about a base on Mars which was made into a movie. This story is his take on a Moon Base:

"Artemis" itself is a five-dome Moon base, servicing a little heavy industry and rather more tourism. Jazz, our heroine, is a sparky young woman who (while her observant Muslim father tut-tuts) gets drunk, has sex and generally tries to have a good time. It's a struggle, though: good times are expensive on the Moon, and despite

supplementing her job – she is a porter – with some judicious smuggling, Jazz is always short of money. She lives in a coffin-sized apartment, shares communal washing facilities and eats the cheapest algae-grown gunk. Poverty persuades her to take on a criminal commission: a little light sabotage on the lunar surface. Naturally, things don't go smoothly: she botches the sabotage, her employer gets murdered, and an assassin is coming after her. The Moon has become a battleground for organized crime over a McGuffin, in this case a new tech that could revolutionize Earth's entire communication system.

6.0 Early Moon Base Design Proposals

Man has been thinking about visiting the Moon since ancient times. The idea of a moon base and living on the Moon is much more recent and as far as I can tell goes back to about 1959 and Project Horizon.

The above drawing shows a lunar base for six to twelve people, built into an inflatable spherical habitat.

Proportions of interior volume devoted to different systems equipment is relatively accurate. The heaviest equipment such as for environmental control, and areas in which the crew spends the most time, such as their personal sleep quarters are lowest in the habitat. Work areas for lunar sample analysis, for hydroponics, and even for small animals are located in the middle areas. The top deck in this view is a running track on which the sloped surface permits the crew member to use centripetal force rather than gravity to permit running in 1/6 G. Concept: NASA (1989)

6.1 Project Horizon

Project Horizon was a 1959 study to determine the feasibility of constructing a scientific / military base on the Moon, at a time when the U.S. Department of the Army, Department of the Navy, and Department of the Air Force had total responsibility for U.S. space program plans. On June 8, 1959, a group at the Army Ballistic Missile Agency (ABMA) produced for the Army a report titled Project Horizon, A U.S. Army Study for the Establishment of a Lunar Military Outpost. The project proposal states the requirements as:

The lunar outpost is required to develop and protect potential United States interests on the Moon; to develop techniques in Moon-based surveillance of the earth and space, in communications relay, and in operations on the surface of the Moon; to serve as a base for exploration of the Moon, for further exploration into space and for military operations on the Moon if required; and to support scientific investigations on the Moon.

The permanent outpost was predicted to be required for national security "as soon as possible", and to cost $6 billion. The projected operational date with twelve soldiers was December 1966.

Horizon never progressed past the feasibility stage, being rejected by President Dwight Eisenhower when primary responsibility for America's space program was transferred to the civilian agency NASA

6.2 Subsurface Moon bases

Building a subsurface Moon base was a concept in the early 1960s. Some suggest building the lunar colony underground, which would give protection from radiation and micrometeoroids. This would also greatly reduce the risk of air leakage, as the colony would be fully sealed from the outside except for a few exits to the surface.

The construction of an underground base would probably be more complex; one of the first machines from Earth might be a remote-controlled excavating machine. Once created, some sort of hardening would be necessary to avoid collapse, possibly a spray-on concrete-like substance made from available materials. A more porous insulating material also made in-situ could then be applied. Rowley & Neudecker have suggested "melt-as-you-go" machines that would leave glassy internal surfaces. Mining methods such as the room and pillar might also be used. Inflatable self-sealing fabric habitats might then be put in place to retain air. Eventually an underground city can be constructed. Farms set up underground would need artificial sunlight. As an alternative to excavating, a lava tube could be covered and insulated, thus solving the problem of radiation exposure. An alternative solution is studied in Europe by students to excavate a habitat in the ice-filled craters of the Moon.

6.3 The LESA Moon Base

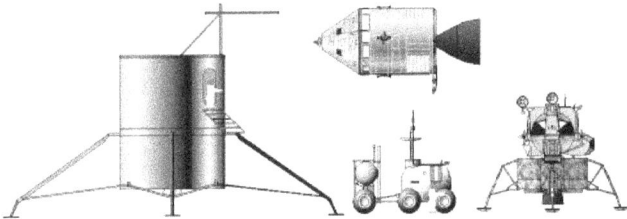

LESA would use a new Lunar Landing Vehicle to land payloads of from 10,500 kg to 25,000 kg on the lunar surface with a single Saturn V launch. Extended CSM and LM Taxi hardware derived from the basic Apollo program would allow crews to be rotated to the ever-expanding, and eventually permanent lunar base. A nuclear reactor would provide power.

Evolution to a lunar base would go from the basic Apollo hardware to AES (Apollo Extension Systems) to ALSS (Apollo Logistics Support System using the LEM Truck), and then ultimately to LESA (Lunar Exploration System for Apollo). Modules developed for ALSS or LEM Truck could be used in LESA systems for commonality and to reduce development costs. The end result would be ever-expanding permanent stations on the Moon.

A typical vision of post-Apollo lunar exploration envisioned the following phases:

2 men/2 days - Apollo

2 men/14 days - AES - LEM Shelter (2050 kg surface payload - LEM Shelter)

2 men/14 to 30 days - ALSS with shelter or MOLAB (4100 kg surface payload)

3 men/90 days - LESA I (10,500 kg surface payload)

3 men/90 days - LESA I + MOLAB (12,500 kg surface payload)

6 men/180 days - LESA II with shelter and extended range roving vehicle (25,000 kg surface payload)

In a comparison of lunar base approaches, the basic Apollo hardware scenario for thorough exploration of a single location would consist of a single manned lunar reconnaissance landing of the selected base site, followed by six Apollo launches over the next six quarters - total, 14 man-days on the Moon for 7 Saturn V launches. The AES or ALSS approach would follow the single reconnaissance flight by three pairs of cargo landings and manned landings, resulting in a total of 86 man-days on the Moon for the same number of Saturn V launches. The LESA approach, with a cargo lander followed by two manned landings in sequence to the same large shelter and rover, would allow 542 man-days on the Moon. ALSS development would cost around $500 million, and LESA cost $1.45 billion. In terms of cost per man-day on the Moon, either approach would pay off on the very first mission.

A diagram of the LESA shelter is below:

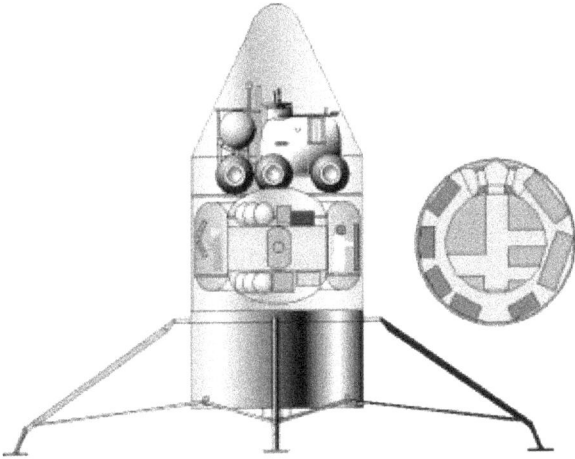

6.4 Underground Base Plan

In 1969 the lunar colony concept (picture below) was developed to encompass a lunar base buried under lunar soil (Johnson, 1969). The sequence of events thought to be possible was a landing in 1969, resources development in 1973-75, a scientific station in 1975, and the lunar colony by 1978.

7.0 The First Moon Landings

Without going into all of the history of the United States manned Moon landings in from 1969 to 1972, it is useful to think about what we learned from these experiences which may apply to future Moon bases.

Here is a map of all of the Moon landings by unmanned probes and manned landers through late 2019:

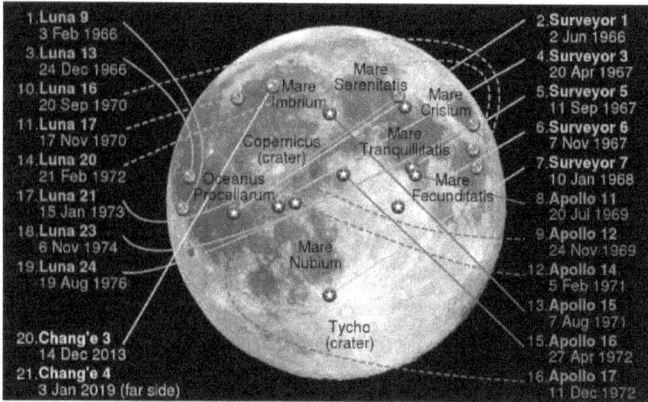

One thing the Moon landings do certainly provide is the confidence that if we got there once we can certainly go there again. Also, with the advancement of technology in fifty years since the first Moon landing we can do a lot more on the Moon than used to be possible.

One of the great limitations of the original manned lunar landings was that they usually had to be within five degrees of the lunar equator. The reason was the limited launch capabilities of one Saturn Five rocket. I know that the Saturn Five is the largest and most powerful rocket built so far, but having more landing flexibility means launching more rockets to carry more materials and fuel to the Moon.

Previous to manned Moon landings some scientists also thought that the Moon would be covered with a sea of Moon dust. This turned out not to be true and a lunar rover was used on later manned flights to travel around the surface.

Since the Apollo landings there have been numerous additional missions to the Moon by unmanned space probes to determine gravity maps, electromagnetic fields, and where water ice is located. Water Ice was mapped to be at both the North and South Poles. (See the picture below)

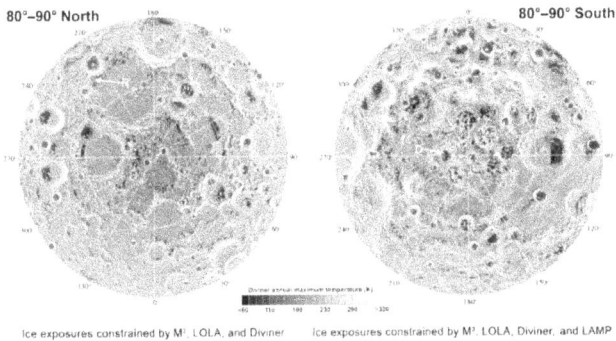

Ice exposures constrained by M^3, LOLA, and Diviner   Ice exposures constrained by M^3, LOLA, Diviner, and LAMP

There are probably also new surprises awaiting us since a Chinese mission to the far side just found some material which acts like a "Gel".

8.0 NASA's Plans for Lunar Travel, Bases & More

The current approach about going to the Moon is to provide an infrastructure to not only land on the Moon but have a transfer station in orbit and build a base on the Moon. Under current directives a Moon base is seen as a learning experience for an eventual manned mission to Mars.

Here is the progression of ideas on the current approach to the Moon supported by NASA:

8.1 Unmanned Lander Plans

An artist's rendering shows Intuitive Machines' Nova-C lander already on the moon. But the 2,240-pound spacecraft isn't scheduled to arrive until next year, when it will be launched by a SpaceX Falcon 9 rocket.

NASA's plan to return to the moon begins with a fleet of robot landers. The space agency has turned to private industry to realize these ambitions, and there are now more than a dozen companies, large and small, building a new generation of robotic lunar landers.

In May 2019, NASA awarded task orders to three of them: Astrobotic Technology, located in Pittsburgh; Edison, New Jersey-based OrbitBeyond, and Intuitive Machines. (OrbitBeyond cancelled its task order two months later, hobbled by financial issues.) Astrobotic plans to deliver 11 payloads to the moon next year. Its destination is Lacus Mortis, a large crater on the near side. Intuitive Machines will also carry packages to the moon next year. It has six payloads going to Oceanus Procellarum, a lunar "sea" that appears as a dark spot on the moon due to its basalt lava surface. In late 2023, Astrobotic will return with another set of payloads—to the Lunar South Pole. Also in 2023, a third vendor, Firefly Aerospace, will carry experiments to "a non-polar region of the moon," and a fourth, Masten Space Systems, is contracted to "deliver and operate" eight payloads to the Lunar South Pole.

The plethora of partners reflects NASA's newly commercial focus. Instead of commissioning spacecraft for companies to build to spec, the space agency is asking industry for mission-ready hardware that can be used by any customer who can pay for the trip. It's a method that has worked for cargo delivery and, albeit with unwelcome delays, regular astronaut trips to the International Space Station.

"For the sake of NASA, we need to be successful," says Altemus. "They won't continue to take those kinds of risks if we fail."

Adding to the pressure, other companies (including billionaire-backed Blue Origin and SpaceX) are leaping into the fray and promising stiff competition on future lunar task orders. It's a cutthroat field, and the competitors are litigious. In August, Blue Origin filed suit against NASA alleging "unlawful and improper evaluation of proposals" in the space agency's selection of SpaceX to build its next lunar lander for astronauts. Meanwhile, China is pressing ahead with a successful rover exploration program, introducing geopolitical gamesmanship and national pride into an already complex landscape.

That puts daily pressure on the IM team to move the project forward, and that leads to long days firing the engines on the abandoned taxiway.

This particular test is happening in January 2020, just eight months after the contract award. (Nine months later, NASA will award IM another contract to deliver a combination drill/mass spectrometer to the moon by the end of 2022.) It takes hours for the IM engineers on the tarmac to ready the test rig for action. They connect power and data lines, set up high resolution video cameras on tripods, endure innumerable dry "click checks" of valves, and finally don protective blue smocks and faceplated helmets to load the cryogenic oxygen.

The pace is as brisk as the January air, but the mood is friendly and collaborative. Everyone chips in on the banal checklist items, like unspooling cables or setting up an awning for shade, regardless of their discipline or rank in the company.

By any metric, this small Houston startup, with around 130 people, is toiling inside an aerospace pressure cooker. So why is everyone here smiling? "We like to do things at a much faster pace than NASA did," says Greg Vajdos, IM's project manager for the Nova-C lander and a former Boeing employee.

The Other Lunar Cycle

After the Apollo program ended, the moon became a forgotten destination. There may have been a lot of talk about a marketplace forming, and even a high-dollar Google X-prize to reach the surface, but since 1976, only China has landed on the moon.

Today it has become a national imperative for the United States to resume lunar exploration. In March 2018, NASA established the Commercial Lunar Payload Services (CLPS) program, creating a fleet of small robotic landers and rovers to scout the moon as a precursor to crewed missions.

It's a necessary forebear of the more ambitious Artemis Program, NASA's plan to establish a long-term human presence on the moon in concert with commercial and international partners, and eventually to use it as a springboard for Mars expeditions. The first step: a lunar lander able to convey 22 pounds of payload to the lunar surface.

When NASA made its lunar intentions clear, Intuitive Machines threw its hat into the ring and pivoted the startup toward a moonshot. The company is a haven, of sorts, for refugees from NASA's Johnson Space Center, where Altemus served as engineering director. In his time at NASA, the agency launched dozens of shuttle flights, from the return to flight after Challenger to the shuttle's retirement. But life at NASA soured for him. "We were trying to create a sense of urgency," says Altemus. "We had the retirement of the space shuttle, we had assembly of ISS completed. Then we had the cancellation of the Constellation Program, and it all became about building capability with no destination. We said, based on that, let's jump."

Intuitive Machines started as a hybrid commercial company/think thank that investigated engineering solutions for the aerospace, energy, and health care industries. The venture is backed by entrepreneur Kam Ghaffarian, co-founder of Stinger Ghaffarian Technologies, NASA's second largest engineering services provider.

But until 2018, flying drones over fires, not landers on the moon, was at the top of the IM agenda. That's when a shift in U.S. policy regarding the moon created a renewed lunar "mission focus and the sense of urgency," Altemus says. The small firm bested others for the first CLPS launch, and the race was on. Staff doubled and Intuitive Machines now has a mix of NASA veterans and fresh-out-of-school engineers.

Mario Maggio wanted to come to Houston, still a beacon for human spaceflight, after he received his master's in aerospace engineering (focus on bioastronautics) at University of Colorado Boulder. Now the 25-year-old is part of a team working on a lunar lander for NASA. "Not too many people who are not retiring will have experience landing on the moon," he says.

University partners on the Nova-C project include Utah State, the University of Colorado, Embry-Riddle Aeronautical University, and Texas A&M. There's also a local community college with an office in the IM Houston facility, a pipeline that turns students into high-tech workshop employees.

The young company has another legacy from its NASA roots: Texas regional pride. "It's our birthright as a city," Altemus says. "If Houston can't field the team to land on the moon, what city can?"

His optimism has survived disappointment. Intuitive Machines partnered with Boeing on a human-rated NASA lander, but Boeing lost the Human Landing Systems (HLS) bidding war among entrants from Blue Origin, SpaceX, and Dynetics. Still, having developed hardware for the Artemis program, the staff at IM believe there's room for them at the table. For example, their 3D-printed, VR3500 lander engine holds the record for continuous test duration at Marshall Space Flight Center's Test Stand 115. It may yet become a part of another outfit's lander. "We expect to be teamed with one of the HLS-contract-winning companies," say Josh Marshall, an IM spokesperson.

Like the rest of the industry, IM is being buoyed by a historic confluence of government funding and commercial potential—none of which means much if IM doesn't stick its landing.

Heartbreak on Regolith

The aeronautical world received a dramatic reminder of how hard it is to land on the moon in April 2019, when a private Israeli organization, SpaceIL, closed in on the final minutes of its lunar journey. The Beresheet lander traveled 3.4 million miles, with just 10 more to go.

The spacecraft started for the surface, using an automated system to fire the main thruster and slow the craft. "The orbital velocity was about 1.7 kilometers per second and needed to be reduced to zero in order to reach the surface softly," recalls Yoav Landsman, the senior systems engineer at SpaceIL during the landing.

"The physics of orbits is that if you reduce your velocity, you lose altitude," he says. "You literally start falling toward the ground at an increasing vertical speed." When it comes to the moon, gravity is the real enemy. "The lack of atmosphere is not the main criterion," Landsman says. "Landing on small asteroids with no atmospheres is more like docking to a space station than landing, since they have a very weak gravity. The moon, on the other hand, is very different because it's massive and pulls hard."

The SpaceIL team uploaded commands to the spacecraft to rotate so the main engine pointed against the direction of the orbital velocity, and then fired the engine. The orbital velocity began to drop, taking almost 15 minutes to zero out. Another pending rotation would bring the spacecraft into its landing orientation, to drop straight down and using the main engine to slow to a near halt.

A telemetry indicator blinked red at 19:22; the lander was plummeting at nearly 75 meters per second. Then the light went dark: malfunction.

The SpaceIL team scrambled to restart the system. In such a scenario, "there is a very slim chance of intervention, because of the communication lag due to the long distance," Landsman says. "You would need to get an indication of the failure, reach a decision, and send a response. If the failure results in an engine cut-off, then you cannot start it in time by command from Earth."

The engine did restart, but too late. Beresheet's vertical speed was 134 meters per second, and its horizontal speed 947 meters per second. At just a mile away from the surface, the lander was doomed. The loss of signal, like a coffin closing, occurred at 19:23.

Landsman says it's a lesson in deep space engineering. "It means that the autonomous process should be very robust and resilient, with a proper onboard redundancy and backup management," he says. There may be another lesson he learned; his skill set could be useful within an emerging lunar landing industry.

"Since there aren't many people in the world who have sent spacecraft to the surface of the moon, I feel that I have a huge responsibility to help human-kind make the next giant leap," he says. "I am currently on the first step of starting my own commercial company to make lunar landings more affordable and accessible."

A dynamic view of the 45th hot-fire test of the IM Nova-C's engine in March 2021. The use of a liquid-methane-powered engine to power a lunar spacecraft will be but one of many firsts IM achieves as commerce leaves its footprints on the moon.

Build, Fly, and Evolve

Rob Morehead, IM's in-space propulsion principal and designer of both main and reaction control system (RCS) engines, sits inside the command trailer and gazes intently at the schematic diagrams on the wide computer screens mounted on the interior trailer wall. Today's tests are focused on a new dual-ignition system.

These are deep-space engines, made to operate autonomously, guided by commands previously uploaded from Earth. The flight-control computer is the ultimate operator of the engine. A gout of greenish-blue flame spears from the thruster's 0.38-inch nozzle, knocking a flock of tiny ice particles from the LOX (liquid oxygen) lines into the vacuum chamber. The chamber is open. These are just ignition tests and as such don't need to replicate a space environment.

That green color in the flame means some metal has burned; the methane/LOX flames are typically a clear blue hue. The next test burns more clearly, and the next creates a darker blue color. The tests are almost indistinguishable, but in each Morehead and his team have made adjustments, seeking the secret recipe of pressure, temperature, and fuel-oxidizer mixture that will set the lander down safely. The speed and strength of the thruster at ignition is vital for RCS engines, which must respond immediately for precise control of the spacecraft.

The use of methane is novel. When Nova-C launches, it will be the first lunar spacecraft to use an engine powered by the reaction of liquid oxygen and liquid methane.

Current vehicles use "hypergolic" fuels like monomethyl hydrazine, which packs a good punch by igniting instantly when combined with nitrogen tetroxide but is also toxic. Though it wasn't Morehead's main reason for settling on methane, one advantage is that it could be manufactured on Mars using components of the atmosphere and subsurface ice.

Engines of this type "have never been flown in space, so this is our chance to prove it's not a risk," says Morehead. The Nova-C's main engine will slow the craft, while the RCS engines govern its orientation. Both need to work precisely to translate the automated system's commands into immediate action during the landing.

"When we go to land on the moon, we want to aim for a field that has a 99 percent chance of landing successfully, in a 300-meter circle," says Altemus. "Then with our landing site initially identified, we screen the area for boulders and rocks that might jeopardize our safe landing."

The other 2022 CLPS lander, Astrobotic's Peregrine, uses a visual system to navigate, but also bounces laser pulses off the surface during its descent to gauge its speed and avoid hazards. The company developed that system under NASA contracts going back to 2014, and tested it in Mojave, California on landers built by the small firm Masten Space Systems. (See "Land Right Here!" Feb./Mar. 2020.)

Because Nova-C uses only visual image processing to land, there are limitations. The same way an optical bombsight doesn't work over water for a lack of landmarks, the dark lunar surface gives the spacecraft no clues on its specific location. "There's certain lighting conditions that we have to have to land on the moon," Altemus says. "We've relaxed the requirement to land anywhere, any time on the moon. Those are requirements that get you in trouble."

Staff with NASA experience gesture in the general direction of the Johnson Center, where "good enough" equated to "unacceptable." Here, it's different, holding to the "new space" ethos of "build, fly, and evolve along the way."

"When I was at NASA, we wrote every single requirement down that we could imagine, before the first procurement went out to any of the vendors, so they could bid it as close as they could," Altemus says. "Why did we write ourselves into those requirements, and then make it so hard to change them, and give the contractor no flexibility? Here, we're able to build a lander that isn't perfect, but can land on the moon. After we've landed on the moon, [we] refine it, over and over and over."

Intuitive Machines plans to fly to the moon with the Nova-C even as they develop the Nova-D, which will have the capacity to carry 1,100 pounds to the lunar surface.

For engineers and executives accustomed to watching this technology develop at a glacial pace, the rapid evolution of both the product and the market is thrilling to witness. "This is the most fun I've had in my career," Altemus says.

While the veterans enjoy the freedom of the commercial shift, the younger staff expect the future to get only brighter. Many of them say they're the generation that will oversee the exploration and utilization of the solar system.

"The moon is an access point to farther destinations," says William Amtmann, a 22-year-old mechanical engineer whose role testing IM's engines is his first full-time job in aerospace. "This is opening up a whole new field."

Lunar Lander Pole Positions

Securing a niche in the lunar marketplace means staying ahead of the rest of the emerging competitors, many of which also hold NASA contracts. The original nine companies with Commercial Lunar Payload Services (CLPS) contracts, which makes them eligible to bid on future NASA task orders, were joined in 2020 by five more.

Everyone involved is hoping that there will be enough variety of payload sizes coming from NASA to give many of them a chance to fly. Astrobotic's shot will come when it carries up NASA's Volatiles

Investigating Polar Exploration Rover (VIPER) to the lunar South Pole for a 100-day mission in late 2023 seeking water ice.

The $199.5 million task order represents a major win for one of the original CLPS competitors. Astrobotic has been in business for 14 years, compared to IM's commercial birth in 2018. The Peregrine lander is on track to hitch a ride with other payloads on the Vulcan Centaur launch system, United Launch Alliance's replacement for Atlas and Delta rockets, in 2022. But delays with the Blue Origin-supplied BE-4 engines might cause the selection of a new launch provider, as happened with the Europa Clipper mission.

Another mission, IM-2, is slated to send the first lander to the moon's South Pole in December 2022. IM's Nova-C will carry the Polar Resourced Ice Mining Experiment-1, the first-ever lunar ice drill, and a rover to demonstrate that 4G communications can work on the moon. IM scored another win with partner Arizona State University in July 2021 by securing a $41.6 million Tipping Point contract from NASA to operate a "hopper-lander" on the moon during IM-2. Deploying from the Nova-C, this Micro-Nova will examine deep lunar craters. Data from the rover's optical imagers will be a boon to any future lunar ice miner.

If private companies, universities, and research organizations can pool their resources, the moon is now in reach. This may help IM's former NASA staffers avoid a repeat of their fate at the space agency, where shifting funding priorities idled them. The ultimate hedge against such reversals of fortune is to fly private lunar missions. IM is doing just that. In August, it announced it would be renting space on its second NASA-sponsored moonshot, IM-2. Four customers will piggyback on IM-2's Falcon 9 rocket.

What's perhaps even more commercially ambitious is IM's bid to dominate the private sector in lunar communications. IM-2's Nova-C will launch a York Space Systems satellite into lunar orbit—to provide bandwidth for use by educational and private sector customers.

Which can't happen until the liquid-oxygen/liquid-methane engines are perfected.

That's what IM is working on this day back in January 2020 at Ellington Field. It's a long day. Homemade sandwiches are extracted from bags. The only bathroom is a bucket with a plastic seat. The main engine never ignites. The staff swarm the vehicle, replacing valves and purging the lines.

It appears the oxygen supply has too much moisture and ice is clogging the valves. The hunt for a new supplier, an unwelcome but small setback, will begin the next day.

The RCS tests fare much better, igniting dozens of times. Each brief burst provides data about the engine's performance and feeds into the software team's landing simulations. The air temperature drops even before the sun sets. A busy and frustrating day has come to an end, but the chore of packing up the gear lies ahead. Cables are respooled and stowed, umbilicals disconnected, the tent broken down in the now whipping wind.

That's the downside of mobile test stand operations, of which there will be dozens more. In October 2021, reached by phone while filling up gas cans for the 57th hot-fire test in two-and-a-half years, program manager Vajdos says IM is building a "flame range." It's a bunker, essentially, that will allow engine test firings on mobile test stands within its walls, without schlepping out to the disused taxiway that was the site of last January's test.

As that January 2020 test draws to a close, the moon clears the horizon and blur of the Earth's atmosphere, and is now gleaming, nearly full, in the early evening sky. The staff's eyes tend to flicker up at it as they work. If they need any motivation, it's beckoning overhead.

8.2 The Deep Space Gateway

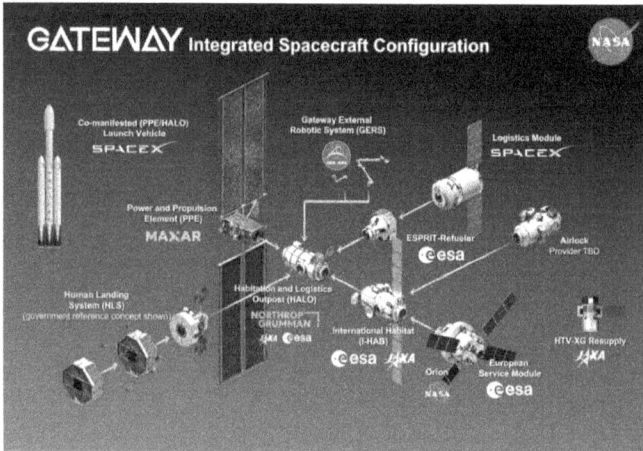

The Gateway will be an international effort and of over twenty five countries who want to participate.

The Lunar Gateway is designed to provide a way station for trips to land on the Moon. It will be in an elongated orbit around the Moon positioned such that landing craft can leave the gateway to land on different parts of the Moon from different orbital positions of the Gateway.

Lunar Orbit Details:

The Lunar Gateway is planned to be deployed in a highly elliptical seven-day near-rectilinear halo orbit (NRHO) around the

Moon, which would bring the station within 3,000 km (1,900 mi) of the lunar north pole at closest approach and as far away as 70,000 km (43,000 mi) over the lunar south pole. Traveling to and from cislunar space (lunar orbit) is intended to develop the knowledge and experience necessary to venture beyond the Moon and into deep space.

The proposed NRHO orbit would allow lunar expeditions from the Gateway to reach a low polar orbit with a delta-v of 730 m/s and a half a day of transit time. Orbital station-keeping would require less than 10 m/s of delta-v per year, and the orbital inclination could be shifted with a relatively small delta-v expenditure, allowing access to most of the lunar surface. Spacecraft launched from Earth would perform a powered flyby of the Moon (delta-v = ~180 m/s) followed by a ~240 m/s delta-V NRHO orbit insertion burn to dock with the Gateway as it approaches the apoapsis point of its orbit. The total travel time would be 5 days; the return to Earth would be similar in terms of trip duration and delta-V requirement if the spacecraft spends 11 days at the Gateway. The crewed mission duration of 21 days and ~840 m/s delta-V are limited by the capabilities of the Orion life support and propulsion systems.

Lunar Gateway Modules are planned as follows:

Contracted modules

The **Power and Propulsion Element (PPE)** started development at the Jet Propulsion Laboratory during the now canceled Asteroid Redirect Mission. The original concept was a robotic, high performance solar electric spacecraft that would retrieve a multi-ton boulder from an asteroid and bring it to lunar orbit for study. When ARM was cancelled, the solar electric propulsion was repurposed for the Gateway. The PPE will allow access to the entire lunar surface and act as a space tug for visiting craft. It will also serve as the command and communications center of the Gateway. The PPE is intended to have a mass of 8-9 tons and the capability to generate 50 kW of solar electric power for its ion thrusters, which can be supplemented by chemical propulsion. It is currently planned to launch on a commercial launch vehicle in 2022. In May 2019, Maxar Technologies was contracted by NASA to manufacture this module, which will also supply the station with electrical power and is based on Maxar's 1300 series satellite bus. The PPE will use Advanced Electric Propulsion System (AEPS) Hall-effect thrusters. Maxar was awarded a firm-fixed price contract of $375 million to build the PPE. NASA is supplying the PPE with an S-band communications system to provide a radio link with nearby vehicles and a passive docking adapter to receive the Gateway's future utilization module.

The **Habitation and Logistics Outpost (HALO)**, also called the Minimal Habitation Module (MHM) and formerly known as the Utilization Module, will be built by Northrop Grumman Innovation Systems (NGIS). A commercial launch vehicle would launch the HALO before the end of year 2023. The HALO is based on a Cygnus Cargo resupply module to the outside of which radial docking ports, body mounted radiators (BMRs), batteries and communications antennae will be added. The HALO will be a scaled-down habitation module, yet, it will feature a functional pressurized volume providing sufficient command, control & data handling capabilities, energy storage and power distribution, thermal control, communications and tracking capabilities, two axial and up to two radial docking ports, stowage volume, environmental control and life support systems to augment the Orion spacecraft and support a crew of four for at least 30 days.

The **European System Providing Refueling, Infrastructure and Telecommunications (ESPRIT) service module** will provide additional xenon and hydrazine capacity, additional communications equipment, and an airlock for science packages. It will have a mass of approximately 4 tons (8,800 lb), and a length of 3.91 m (12.8 ft). The studies and design are being performed mostly by Airbus and OHB. The module construction was approved in November 2019.

The **International Habitation Module (iHAB)** will be an additional habitation module built by ESA in collaboration with Japan. Together with the HALO module, they will provide a combined 125 m3 (4,400 cu ft) of habitable volume to the station.

Proposed modules

The concept for the Lunar Gateway is still evolving, and these modules have also been proposed to be added to the design:

The **Gateway Logistics Modules** will be used to refuel, resupply and provide logistics on board the space station. The first logistics module sent to the Gateway will also arrive with a robotic arm, which will be built by the Canadian Space Agency.

The **Gateway Airlock Module** will be used for performing extravehicular activities outside the space station and would have the docking port for the proposed Deep Space Transport.

8.3 Modern Lunar Lander Concepts

NASA Originally picked SpaceX as their lunar lander choice based on the following specifications. Then recently they decided to contract for a second lander design. The press release for that is later in this chapter.

Lunar landers for the next phase of Moon landings will not be one time landers but reusable vehicles.

In October 2020, NASA officials wants to select two contractors from the design study award winners. Those firms will proceed with full development of their human-rated lunar landers, and NASA will later choose one for a landing attempt in 2024, and another for a Moon landing in 2025.

NASA officials want to conduct lunar landing missions on a cadence of at least one per year after 2024.

Two astronauts are expected to be aboard for the Artemis program's first two landing attempts, including — as NASA regularly mentions — the first woman to land on the Moon.

Among other requirements, the landers for the 2024 and 2025 missions must provide the following capabilities:

At least 1,907 pounds (865 kilograms) of payload delivered to the lunar surface, with a goal of 2,127 pounds (965 kilograms)

At least 6.5 days on the lunar surface

At least two spacewalks per mission, with goal of five spacewalks, using NASA-provided spacesuits

At least 77 pounds (35 kilograms) of sample return capability, with a goal of 220 pounds (100 kilograms)

NASA will pick one or both of the lunar lander developers to work on a more "sustainable" lander design, a craft that will be able to carry at least four astronauts to the Moon's surface, operate during the two-week-long lunar night, and support longer-duration spacewalks. The more advanced lander could be ready to fly to the Moon in 2026, and must utilize NASA's Gateway in lunar orbit.

Latest 2022 press release on a Lunar Lander design

Today, NASA announced plans to develop a second human lunar lander for its Artemis program, the agency's major spaceflight initiative to send humans back to the Moon. To build the vehicle, the space agency is calling on commercial space companies to propose concepts for landers that can take people to and from the Moon's orbit and the lunar surface, with the goal of having them ready by 2026 or 2027 at the earliest.

NASA already holds a contract with commercial partner SpaceX to develop a lunar lander for Artemis, which aims to land the first woman and the first person of color on the Moon. In 2021, the space agency awarded a sole contract to SpaceX worth $2.9 billion to develop the company's future Starship vehicle into a lander that can take humans to and from the lunar surface. As of now, both NASA and SpaceX are working toward conducting the first Artemis lunar landing as early as 2025, though that timeline is considered unlikely.

NASA HAD ORIGINALLY WANTED TO PICK TWO COMPANIES TO DEVELOP HUMAN LUNAR LANDERS

NASA had originally wanted to pick two companies to develop human lunar landers for Artemis in order to inspire competition and keep down costs. The agency was going to pick the two winners from three finalists: SpaceX, Blue Origin, and Dynetics. But the agency ultimately chose one, primarily due to budget constraints. For the year 2021, NASA had requested $3.4 billion from Congress to fund the development of Artemis lunar landers but only received $850 million, just 25 percent of what was asked. As a result, NASA went with SpaceX, in part because the company had offered the most affordable bid.

However, the decision to simply pick one company didn't sit well with the losing finalists. Blue Origin proceeded to sue NASA in federal court over the selection, though the company ultimately lost its case. Despite the lawsuit, NASA administrator Bill Nelson expressed his desire to eventually have two lunar lander providers, with hopes that Congress would fund the initiative. And at one point, it looked as if Congress would direct NASA to make that happen. In October, the Senate Appropriation Committee introduced a bill that would direct NASA to pick a second company to develop a lunar lander for Artemis. However, the most recent budget bill that was signed for 2022 did not force NASA to do that, but it did give the space agency the full $1.195 billion it asked for to develop lunar landers.

"I PROMISED COMPETITION, SO HERE IT IS."

Now, ahead of President Joe Biden's budget request expected next week, NASA is announcing official plans to select another company's lunar lander, as the agency had wanted to do all along. "Competition leads to better, more reliable outcomes," Nelson said during a press conference announcing the news. "It benefits everybody. It benefits NASA. It benefits the American people." Nelson added: "I promised competition, so here it is." NASA now plans to put out a draft call for proposals at the end of the month, with plans to release a final call later this spring. Everyone but SpaceX will be able to compete in this new competition for a contract.

In light of this announcement, NASA says it will make some changes to SpaceX's existing contract. NASA's original contract with the company has SpaceX conducting an uncrewed landing on the Moon as a demonstration test before conducting the first crewed landing in the mid-2020s. That first landing, which will mark NASA's return to the Moon with humans since the 1970s, will be called Artemis III and should receive plenty of attention. After that, SpaceX would have moved on to a new operational contract with NASA, where the space agency would buy individual flights of the Starship lander to continue going back to the Moon.

Now, NASA plans to work with SpaceX under its original contract to conduct a possible third crewed landing after Artemis III. Then after that, SpaceX and the new company that NASA picks would presumably compete for upcoming Artemis missions — that is, if the providers are ready.

A lot of this depends on buy-in from Congress, which may happen since some lawmakers indicated their desire for NASA to pick a second lunar lander. NASA would not say how much it plans to ask for in development funding, but with President Biden's budget request scheduled for release on March 28th, the numbers should be available fairly soon. Nelson says he expects the funds to materialize. "We're expecting to have both Congress support and that of the Biden administration, and we're expecting to get this competition started in the fiscal year 2023 budget."

8.4 SpaceX's Starship Lunar Lander

NASA has announced that SpaceX will have competition when it comes to putting a commercially operated crewed spacecraft on the surface of the moon as part of the Artemis mission.

The space agency announced on Wednesday it is asking other American companies to propose lunar lander concepts capable of taking astronauts from lunar orbit to the moon's surface.

The proposals will form part of the Artemis missions after Artemis III, currently set to land its first astronauts on the moon in 50 years.

The second commercial contract, known as the Sustaining Lunar Development contract, represents a second pathway for lunar lander development alongside the ongoing work with Elon Musk's SpaceX.

NASA said the effort is meant to maximize NASA's support for competition and provides redundancy in services to help ensure NASA's ability to transport astronauts to the lunar surface.

SpaceX was selected by NASA as a crewed lunar-landing partner in April 2021. The company's lunar lander is known as Starship, and it is expected to travel to the lunar surface no sooner than 2025.

As part of the Artemis mission, will see the first person of color and the first woman step foot on the surface of the moon. NASA has also allotted a further mission to SpaceX's Starship spacecraft along with those already planned.

Not only will Starship ferry crew to the moon, but Musk's SpaceX says it will also transport large amounts of cargo that will form the building blocks of future space exploration research and human spaceflight development.

The fully reusable Starship spacecraft and its Super Heavy rocket launch system is predicted to be the most powerful launch vehicle ever developed, capable of lifting cargo exceeding 100 metric tons into orbit around Earth.

Starship is powered by three reusable methalox staged-combustion Raptor engines, which began flight testing on the prototype rocket in July 2019.

This lifting power and cargo-hauling capability will improve sustainability in space, allowing for longer space missions and enabling the moon to be used as a stepping stone to the exploration of Mars. Starship is also planned to travel to the red planet multiple times, with its heat shield designed to withstand multiple entries into the Martian atmosphere.

The new commercial spacecraft which will join Starship in journeying to the moon, and will be built in accordance with NASA's long-term lunar landing requirements.

It will have to possess the capability to dock at a planned lunar orbiting space station known as Gateway, increase crew capacity, and have the capacity to transport more science and technology to the surface.

NASA Administrator Bill Nelson said in a statement to the press: "Under Artemis, NASA will carry out a series of groundbreaking missions on and around the Moon to prepare for the next giant leap for humanity: a crewed mission to Mars.

"Competition is critical to our success on the lunar surface and beyond, ensuring we have the capability to carry out a cadence of missions over the next decade. Thank you to the Biden Administration and Congress for their support of this new astronaut lander opportunity, which will ultimately strengthen and increase flexibility for Artemis."

8.5 Settling the South Pole

Most current proposals for settling the Moon are focused on Shakelton Crater at the Lunar South Pole. Why? Because previous sensing satellites have found a large amount of Water ice at this location. The hope is that water ice can be minded and refined to provide water for habitation and decomposed for rocket fuel.

8.6 Far Side Bases

There are some good reasons to build bases on the far side of the Moon which never faces the Earth. One is radio astronomy which would be protected from all of Earth's electromagnetic emissions. Another reason are large telescopes. Since the Moon doesn't have any atmosphere they can be built on the far side which is both very dark and with no atmospheric distortion of telescopic images. These images would be incredible and since it will be on the Moon it can be serviced and improved as needed.

9.0 Types of Moon Bases

There are different ideas for building a Moon Base which all have positive and negative benefits and detractions.

Here are some of the major concepts below:

9.1 Using Lava Tubes

Lunar lava tubes are ancient volcanic tunnels on the Moon that are thought to have formed during basaltic lava flows. When the surface of a lava tube cools, it forms a hardened lid that contains the ongoing lava flow beneath the surface in a conduit-shaped passage. Once the flow of lava diminishes, the tunnel may drain, forming a hollow void. Lunar lava tubes are formed on surfaces that have a slope that ranges in angle from 0.4° to 6.5°. Lunar lava tubes may be as wide as 500 meters (1,600 ft) before they become unstable against gravitational collapse. However, stable tubes may still be disrupted by seismic events or meteoroid bombardment.

The existence of a lava tube is sometimes revealed by the presence of a "skylight", a place in which the roof of the tube has collapsed, leaving a circular hole that can be observed by lunar orbiters. Here are some pictures of possible lava tubes on the Moon:

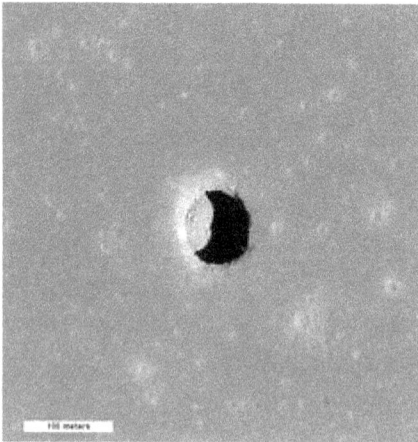

The next picture shows a possible tunnel on the Moon:

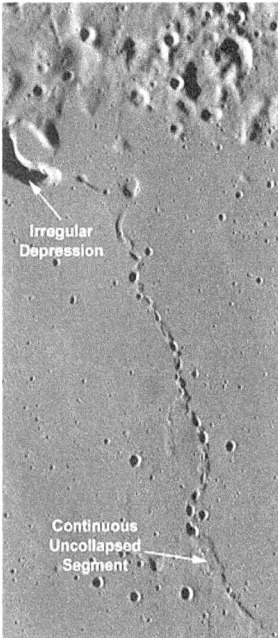

Using manmade tunnels and natural lava tubes for a structure has the benefits of good protection from solar radiation and making it easier to trap an atmosphere by just enclosing two sections of the tunnel. The problems have to do with the equipment needed for any tunnel digging needed and having to find a lava tube to build a base which might be in an inconvenient location.

9.2 Printing Buildings

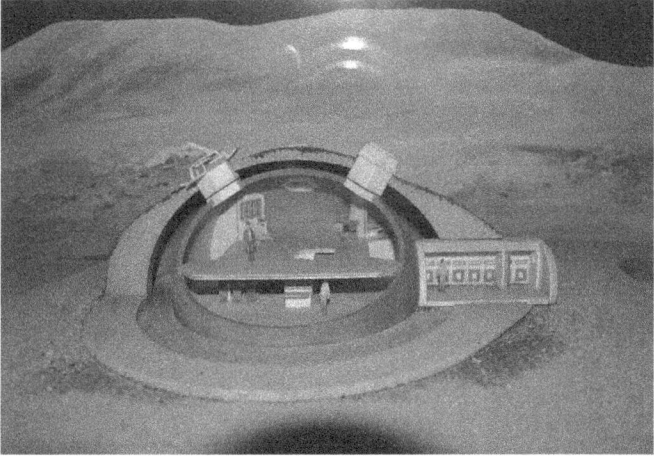

Three dimensional printers were developed from the concept of ink jet paper printers. What if you could use other materials than ink to deposit and build up an object layer by layer from a computer design?

Three D printers are now used widely around the world and are being used to print even complex objects like rocket engines.

Some people have even experimented with using this approach to build homes or other buildings on Earth.

This had led some Moon Colony designers to think about just using Lunar Regolith and a printing robot to build an entire shelter or pile up and compress regolith over a balloon structure. Maybe a combination of the two concepts will lead to the least expensive and quick to build lunar base.

9.3 Underground Structures

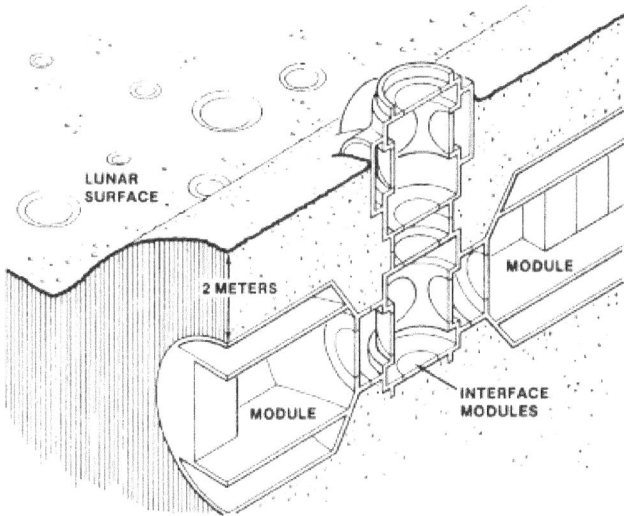

Building a structure underground has the benefits of radiation protection and makes it easier to contain an atmosphere. The problem is that it takes a lot of digging equipment to either dig pits for module structures or to dig tunnels for those structures. This is a more expensive approach to building an early Moon base than building on the surface. However, underground structures do have the benefit of being safe from radiation and easier to control atmospheric losses.

10.0 Lunar Space Suits

The original Apollo Space and Moon suit design had a lot of complexity:

Specifications, Apollo 7 - 14 EMU

Name: Extravehicular Mobility Unit (EMU)

Manufacturer: ILC Dover (Pressure Suit Assembly) and Hamilton Standard (Portable Life Support System)

Missions: Apollo 7-14

Function: Intra-vehicular activity (IVA), orbital Extra-vehicular activity (EVA), and terrestrial EVA

Operating Pressure: 3.7 psi (25.5 kPa)

IVA Suit Mass: 62 lb (28.1 kg)

EVA Suit Mass: 76 lb (34.5 kg)

Total EVA Suit Mass: 200 lb (91 kg)

Primary Life Support: 6 hours

Backup Life Support: 30 minutes

Extravehicular Pressure Suit Assembly

Figure 1-5. - Extravehicular pressure garment assembly with arm bearing

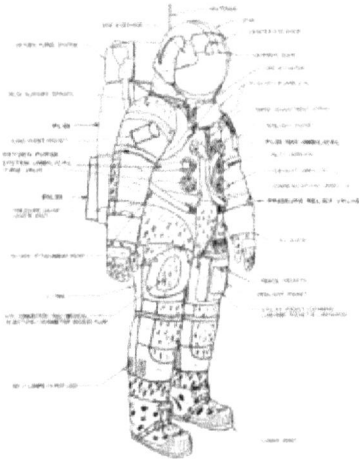

Figure 1-1. - Extravehicular mobility unit with LJTMG.

Torso Limb Suit Assembly

Between Apollo 7 the Commander (CDR) and Lunar Module pilot (LMP), had Torso Limb Suit Assemblies (TSLA) with six life support connections placed in two parallel columns on the chest. The 4 lower connectors passed oxygen, an electrical headset/biomed connector was on the upper right, and a bidirectional cooling water connector was on the upper left.

Integrated Thermal Micrometeoroid Garment

Covering the Torso Limb Suit Assembly was an Integrated Thermal Micrometeoroid Garment (ITMG). This garment protected the suit from abrasion and protected the astronaut from thermal solar radiation and micrometeoroids which could puncture the suit.

The garment was made from thirteen layers of material which were (from inside to outside): rubber coated nylon, 5 layers of aluminized Mylar, 4 layers of nonwoven Dacron, 2 layers of aluminized Kapton film/Beta marquisette laminate, and Teflon coated Beta filament cloth.

Additionally, the ITMG also used a patch of 'Chromel-R' woven nickel-chrome (the familiar silver-colored patch seen especially on the suits worn by the Apollo 11 crew) for abrasion protection from the Portable Life Support System (PLSS) backpack. Chromel-R was also used on the uppers of the lunar boots and on the EVA gloves. Finally, patches of Teflon were used for additional abrasion protection on the knees, waist and shoulders of the ITMG.

Starting with Apollo 13, a red band of Beta cloth was added to the commander's ITMG on each arm and leg, as well as a red stripe on the newly added EVA central visor assembly. The stripes, initially known as "Public Affairs stripes" but quickly renamed "commander's stripes", made it easy to distinguish the two astronauts on the lunar surface and were added by Brian Duff, head of Public Affairs at the Manned Spacecraft Center, to resolve the problem for the media as well as NASA of identifying astronauts in photographs.

Liquid Cooling Garment

Lunar crews also wore a three-layer Liquid Cooling and Ventilation Garment (LCG) or "union suit" with plastic tubing which circulated water to cool the astronaut down, minimizing sweating and fogging of the suit helmet. Water was supplied to the LCG from the PLSS backpack, where the circulating water was cooled to a constant comfortable temperature by a sublimator.

Portable Life Support System

At the beginning of the Apollo spacesuit competition, no one knew how the life support would attach to the suit, how the controls needed to be arranged, or what amount of life support was needed. What was known was that in ten months, the Portable Life Support System, aka "backpack", needed to be completed to support complete suit-system testing before the end of the twelfth month. Before the spacesuit contract was awarded, the requirement for normal life support per hour almost doubled. At this point, a maximum hourly metabolic energy expenditure requirement was added, which was over three times the original requirement.

In late 1962, testing of an early training suit raised concerns about life support requirements. The concerns were dismissed because the forthcoming Apollo new-designs were expected to have lower effort mobility and improved ventilation systems. However, Hamilton took this as a strong indication that Apollo spacesuit life support requirements might significantly increase and initiated internally funded research and development in "backpack" technologies.

In the tenth month, the first backpack was completed. Manned testing found the backpack to meet requirements. This would have been a great success but for the crewed testing confirming that the 1963 life support requirements were not sufficient to meet lunar mission needs. Early in 1964, the final Apollo spacesuit specifications were established that increased normal operations by 29% and increased maximum use support 25%. Again, the volume and weight constraints did not change. These final increases required operational efficiencies that spawned the invention of the porous plate sublimator and the Apollo liquid cooling garment.

The porous plate sublimator had a metal plate with microscopic pores sized just right so that if the water flowing under the plate warmed to more than a user-comfortable level, frozen water in the plate would thaw, flow through the plate, and boil to the vacuum of space, taking away heat in the process. Once the water under the plate cooled to a user-comfortable temperature, the water in the plate would re-freeze, sealing the plate and stopping the cooling process. Thus, heat rejection with automatic temperature control was accomplished with no sensors or moving parts to malfunction.

The Apollo liquid cooling garment was an open mesh garment with attached tubes to allow cooling water to circulate around the body to remove excess body heat when needed. The garment held the tubes against the body for highly efficient heat removal. The open mesh allowed air circulation over the body to remove humidity and additionally remove body heat. In 1966, NASA bought the rights to the liquid cooling garment to allow all organizations access to this technology.

Before the first Apollo spacewalk, the backpack gained a front-mounted display and control unit named the remote control unit. This was revised for Apollo 11 to additionally provide camera attachment to provide high quality lunar pictures.

Modern Moon Suit Designs

New space suits are being designed since the Apollo Moon suit design is over fifty years old.

The two spacesuit prototypes which NASA showcased are designed for two separate parts of a crewed mission to the Moon. One, called the Exploration Extravehicular Mobility Unit (xEMU) is a red, white and blue suit designed to be worn by astronauts exploring the lunar surface, specifically at the Moon's South Pole — the target for NASA's next crewed lunar landing.

The second suit is the Orion Crew Survival System, which is a bright orange pressure suit that will be worn by astronauts when they launch into space on the Orion capsule and return to Earth.

A Recent NASA Spacesuit Announcement

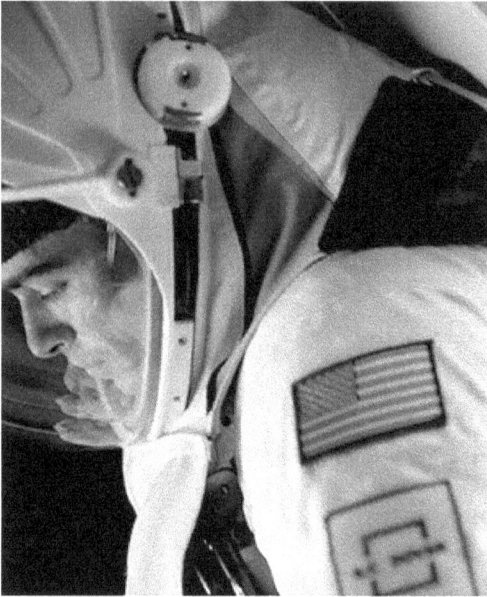

NASA picks two companies to design next moon- and space-suits

June 1, 2022 — When the next American astronauts walk on the moon, they will do so wearing spacesuits furnished either by the company behind the historic Apollo lunar garments or the firm building a commercial outpost to be tested at the International Space Station.

NASA on Wednesday (June 1) selected Collins Aerospace and Axiom Space to develop, build and operate the next generation of U.S. spacesuits and spacewalk systems to be used outside of the space station and on Artemis missions to the moon. The two companies will vie for missions under the agency's Extravehicular Activity Services (xEVAS) contract, which has a potential total value of $3.5 billion to be awarded over the next 12 years.

"This is a historic day for us," Vanessa Wyche, director of NASA's Johnson Space Center in Houston, said in a press conference on Wednesday. "History will be made with these suits. When we get to the moon. we will have our first person of color and our first woman who will wear and use these suits in space, so this is a very important day."

The first task orders to be competed under the xEVAS contract will include the development and services for the first demonstration outside the space station and the Artemis III mission, which is planned to be the first human lunar landing since the last Apollo mission touched down 50 years ago.

Collins Aerospace, a Raytheon Technologies business, has partnered with ILC Dover and Oceaneering to develop its xEVAS spacesuit. NASA's Apollo spacesuit and the Extravehicular Mobility Unit (EMU) that is still in use on the space station today were developed by the Hamilton Sundstrand division of Collins (then Hamilton Standard) and ILC.

"Our heritage corporations ... designed and developed and fielded and operated the Apollo suit, so we've got some surface experience. We've also got more than four decades with a spacesuit design for the space shuttle that has evolved and in its current iteration has been supported for six years on board the space station with only crew members working on it. In the process of doing that we've learned an awful lot," said Dan Burbank, senior technical fellow at Collins Aerospace and a retired NASA astronaut.

"So the goal is to take the foundations that NASA has laid ... and evolve that technology and create a suit that is compatible with the entire spectrum of crew members," said Burbank. "So that commander of Artemis III, she has got a suit that is appropriately sized and tailored for her that doesn't feel like a spacecraft but feels like a ruggedized set of extreme sport outerwear."

Axiom Space, which earlier this year launched the first all-private crew to the space station and holds a NASA contract to dock and test out components for its own private orbital outpost at the ISS, has already been at work developing a spacesuit, the AxEMU, to support its own needs.

"We are building a space station, so we have a need for a spacesuit. We have a number of customers who already would like to do a spacewalk and we had planned to build it soon as part of our program," said Michael Suffredini, president and chief executive officer for Axiom Space and NASA's former space station program manager. "So it's fantastic to have a partnership where we can benefit from the years of experience that NASA has and all the work they've done to advance a design to where it is today and then, as a commercial company, come in and work with them to build it in a way that's lowest cost so that we can both utilize the suit to meet our needs."

Axiom Space's partners for the AxEMU include KBR, Air-Lock, David Clark Company, Paragon Space Development Corporation, Sophic Synergistics and A-P-T Research.

NASA has defined the technical and safety standards by which the spacesuits will be built, but Collins and Axiom will own the spacesuits and will be responsible for their certification and production, along with the equipment needed to support space station and Artemis missions. This approach to spacewalk services is intended to encourage an emerging commercial market for a range of customers.

For the past several years, work has been underway at Johnson Space Center to develop a spacesuit known as the Exploration Extravehicular Mobility Unit, or xEMU. Work on that in-house design will continue through the end of this fiscal year, in parallel with the commercial procurement, before the team migrates to an oversight and collaborative role with Collins and Axiom.

"They're putting it through a lot of testing. In fact, the life support system is in a chamber as we speak, running through a test," said Lara Kearney, manager of NASA Johnson's Extravehicular Activity and Human Surface Mobility Program. "We're making all of the results from the NASA tests available to these companies so they can use any and all of it as they choose."

NASA designed the xEVAS contract so it can evolve as the needs of the agency and space industry change. The contract also provides NASA with the option to add additional vendors that were not selected as the commercial space services market matures.

11.0 Lunar Transportation

We're going back to the moon and General Motors is coming along, too. Just as it did with Boeing back in the 1960s, GM is teaming up with the aerospace experts at Lockheed Martin for the newest versions of the moon buggy. This time, however, much of the technology used in these buggies are coming from the same Ultium EV tech used in the GMC Hummer and Chevrolet Silverado EV trucks.

The Original Buggy

For now, these are the renderings of the future lunar rovers. Unlike the original moon buggy—officially known as lunar rover vehicles or LRVs—these are much larger and will potentially drive farther thanks to modern battery technology.

The original rover was launched in 1971 and created to extend the Apollo missions beyond the walking distance of the astronauts. It was also developed in about 17 months by GM Defense Research Labs, Delco Electronics, and Boeing and built in Kent, Washington. The buggy used four brushed DC motors capable of about 0.25 hp. Only intended push the LRV to 8 mph, astronaut Eugene Cernan was able to push his buggy to 11 mph during the Apollo 17 mission.

Power for the lunar rover was provided by a pair of 36-volt, 4.4-kWh silver zinc batteries that were connected in parallel for a total capacity of 8.8 kWh. While a far cry from our current generation of Lithium-based batteries, the buggy was able to drive as far as 50 miles on a full charge. During Apollo 15 and 16, due to safety concerns, the buggies were nevertheless limited to just 10 miles in a single direction and the initial drive during Apollo 15 was 17.25 miles total. However, by Apollo 17, these concerns were relaxed and that mission's LRV went a total of 22.3 miles with a 12.5-mile long, one-way distance.

The New Buggies

Fortunately, technology has taken some great leaps and bounds since Apollo 17. According to GM back in May of 2021, these next-generation lunar rover vehicles are being designed to drive even farther distances to support the first excursions of the Moon's South Pole. These new LRVs will also need better technology beyond their capacities as the south pole is dark, making it much colder with much more rugged terrain and the sunlit surfaces the Apollo missions landed in.

Fortunately, GM's Ultium technologies will most likely help create the drive and battery packs these new rovers need. These new rovers will also be autonomous, allowing them to launch prior to the human landings. This is needed to help prepare for "commercial payload services and enhance the range and utility of scientific payloads and experiments" according to GM.

Not The Only Player In Town

Another difference from the original Apollo missions, even the ones prior to the rover missions, there will be several lunar missions sanctioned by NASA. One of those missions also includes another rover built by Northrop Grumman. It is joined by AVL, Intuitive Machines, Lunar Outpost, and Michelin to create the Lunar Terrain Vehicle—the LTV. Each brand is bringing its own expertise to the LTV. Northrop Grumman will lead systems integration, bridging its own flight-proven experience with spacecraft design to include cargo storage, energy management, avionics, navigation, sensors, controls, mission planning, operations and training..

AVL is billed as an industry leader for the development, simulation, and testing of vehicle systems and will be used for its expertise in the advancement of battery electric vehicles, autonomous driving, and propulsion solutions. Intuitive Machines and their Nova-D spacecraft will be used to bring the LTV to the Moon using four liquid methane/oxygen engines from their Nova-C program. Lunar Outpost—a leader in lunar mobility platforms according to Northrop—will leverage its dust mitigation and thermal technologies from the development of its MAPP rover to help develop similar solutions for the LTV.

Tweels On The Rover Go 'Round And 'Round

Last, but not least, Michelin is using its own experiences with NASA's expertise in high-tech materials, and its development of airless solutions for extreme applications to design an airless tire for the LTV. Yes, the Tweel is going to the Moon and a lunar rover will probably not have a tire made of zinc-coated piano wire and aluminum strips this time.

We're in an exciting time of space exploration and it looks as if we're finally going back to the moon with some mobile support. It just looks like the technology flow is going the other way, this time.

Instead of what we learned in development of going to the Moon, we're using what we've developed on Earth for everyday travel to go back.

12.0 Lunar Power Production

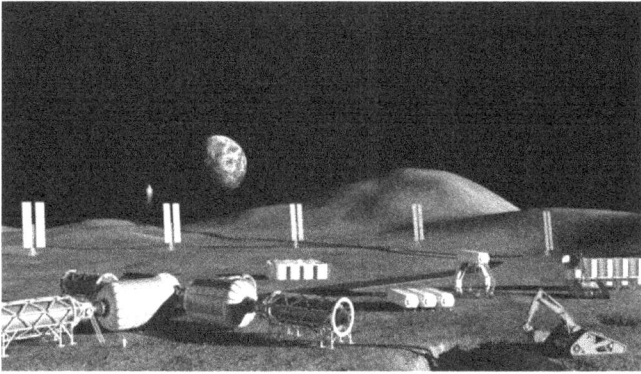

The preferred system recommended in the 2009 NASA study was a photovoltaic solar array-powered cryogenic storage regenerating fuel cell system. NASA calculated that a five-kilowatt continuous delivery system would store 2,000 kilowatt-hours with a system energy density of 1.15 kilowatt-hours per kilogram. The study's alternate preferred system was a fixed orbit laser system, with a 16.1-hour orbit period that required a surface receiver installation with 525 kilowatt-hours of energy storage. The laser was powered and fired when it was both in direct sunlight and in direct line-of-sight with the Moon base.

Nuclear Fission Power is also being Considered

Here is a recent article on the status of these systems:

Space reactors in the kilowatt class are designated **KiloPower** by NASA, and may include a variety of designs of comparable power and mass to RTGs. They use liquid metal heatpipes to transfer fission heat to either thermoelectric or Stirling power conversion. Los Alamos National Laboratory and NASA Glenn Research Center completed a proof-of-concept test at the Nevada National Security Site in 2012 using the Flattop reactor and two small Stirling convertors to produce 24 watts.

In December 2014 NASA's Glenn Centre announced progress with its 4 kWt/1 kWe KiloPower project, using high-enriched uranium powering a heatpipe system and Stirling engine to generate electricity – Kilopower Reactor Using Stirling Technology (KRUSTY). This is a fast reactor relying entirely on negative thermal feedback for control, the objective being to design self-regulation as a major feature and demonstrate that it is reliable. The design is scalable up to 10 kWe. Los Alamos Nuclear Laboratory (LANL) with NASA in April 2018 announced completion of full-power testing of a prototype KRUSTY unit.

The testing ran from November 2017 to March 2018, during which time the unit successfully handled multiple simulated failures, including power reduction, failed engines and failed heat pipes. This was the first US nuclear-powered ground test on an in-space nuclear reactor for several decades.

Prior to testing, NASA had appealed to the US National Nuclear Security Administration (NNSA) to let it proceed. The testing is being carried out under the Department of Energy's Criticality Safety Program working with NASA. The optimum fuel for the fast reactor was proposed as an HEU (93% enriched) alloy with 7% molybdenum as a solid casting, 129 mm in diameter and 300 mm long*. A 250 mm diameter beryllium oxide reflector would surround this, with 18 sodium heatpipes between the fuel and the reflector. Criticality is achieved by raising the BeO reflector to generate fission in the reactor core. Once fission has begun, the BeO reflector would be slowly raised to increase the temperature in the system to 800°C. There is a single central boron carbide control rod. The heatpipes will deliver 13 kWt heat from the core to eight free-piston Stirling engines and allow each to produce about 125 watts of electric power. The Stirling engine would have a cylindrical radiator of nearly 10 m². The system mass is about 750 kg, and the length about 5 m. The science payload is assumed to be about 10 m from the core, and shielded by 45 kg of depleted uranium and 40 kg of lithium hydride.

* The 1 kWe test reactor in November 2017 is reported to use an alloy of 92% uranium, 8% molybdenum, with enrichment to 95%, and diameter 11 cm with 4 cm central hole and eight heat pipes recessed. There are two axial neutron reflectors and one radial one, total 70.5 kg beryllium.

NASA estimates that about 40 kWe would be needed for power on Mars, using ten 4 kWe KiloPower units.

Experience of the KiloPower project will be fed to a **MegaPower** project, with 2 MWe units. Features would include reactor self-regulation, low reactor core power density and the use of heatpipes for reactor core heat removal. The reactor would be attached to an open air Brayton cycle power conversion system using air as the working fluid and as the means of ultimate heat removal. The reactor would weigh about 40 tons including 3 t of LEU fuel (16-19% enriched), and be 4 m long, 2 m diameter. It would be scalable to 10 MWe, and could also be used in military bases, with 72-hour installation.

In April 2021 the US Defense Advanced Research Projects Agency (DARPA) awarded contracts for the first phase of its Demonstration Rocket for Agile Cislunar Operations (DRACO) program to demonstrate a nuclear thermal propulsion (NTP) system above low Earth orbit in 2025. General Atomics will carry out reactor development work, while Blue Origin and Lockheed Martin will develop spacecraft concept designs.

13.0 Communications

Studies done by NASA in 2007 suggest that there be a lunar orbiting communication or satellite or multiple satellites to provide a relay for communications with the Lunar Gateway and Earth. This would also be a digital network to provide IP internet messaging capabilities. Updated studies still need to be done but this one provides a good first approach to providing communications.

The proposed Andromeda satellite constellation is composed of 24 satellites divided evenly among four different orbits in order to provide maximum coverage for the moon's surface. Each satellite in the proposed Andromeda constellation has three different antennas to establish communications with both Earth and the lunar surface. When stored, the entire satellite is 44 by 40 by 37 centimeters.

EARTH'S MOON IS the target of more missions than at any time since the Apollo era, by both space agencies and commercial entities. NASA, for example, has plans to visit the moon using both robots and humans, and is also considering—with international collaborators—a small orbiting outpost in the next decade. This facility, known as the Lunar Gateway, would store supplies, host visiting astronauts, and facilitate communication between the moon and Earth.

Although the gateway is perhaps the most ambitious of the projects planned, it's only one of over 90 lunar missions being considered for the years between now and 2030. Of course, not all these planned missions will materialize, but many—if not most—will happen in some form. And this is only the beginning: We anticipate that interest in the moon will accelerate, eventually culminating in a permanent human presence on the surface.

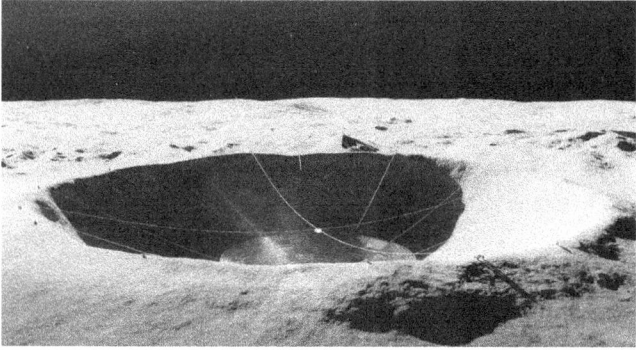

The proposed Lunar Crater Radio Telescope would turn a crater on the far side of the moon into a massive dish-shaped antenna to survey the universe, accumulating massive amounts of data that need to be sent back to Earth for analysis.

If that comes to pass, lunar denizens will need to stay in touch with Earth. While direct radio communication with Earth was used during the Apollo missions, it doesn't work in every possible situation. For example, the moon's far side, as well as large portions of its poles, have no direct line of sight to Earth. Even on the side facing Earth, hills and crater walls can block communications.

And on the practical side, direct communication across hundreds of thousands of kilometers of space requires a powerful communications terminal with a large antenna or a high-wattage amplifier, if not both. Small robots, for example, will not have the space or the power for these large systems. A better solution to lunar connectivity is a network of relay spacecraft orbiting the moon to provide continuous coverage everywhere.

Italian aerospace company Argotec and NASA's Jet Propulsion Laboratory (JPL) are collaborating on the concept of an orbiting relay satellite constellation called Andromeda. Argotec (at which Balossino is head of the R&D unit) is developing spacecraft concepts and JPL (at which Davarian is a project manager) is providing subsystems such as radios and antennas. The approach consists of 24 relay satellites to be placed in a constellation using 4 orbits, with 6 satellites per orbit. This configuration would provide continuous coverage to the poles, and near-continuous coverage everywhere else, with only occasional slight gaps. With this relay system, missions anywhere on the lunar surface would have reliable, consistent connections to Earth.

Placing relay satellites in orbit around the moon comes with challenges. First, we would like to use orbits that are stable—meaning satellites would require little or no maneuvering. Second, orbits need to be selected with continuous or near-continuous physical line of sight to "hot spots" that will likely have considerable human or robotic activity. And third, while guaranteeing high visibility for lunar hot spots, we don't want to deny connectivity to any other portions of the surface as a result.

Any relay-satellite network needs to provide the best possible service and coverage with the minimum number of satellites.

The moon's South Pole is one probable hot spot because its craters contain ice, at least to some extent. For longer crewed missions, the water that humans require would likely be easier to harvest from the moon rather than to haul it from Earth. Water can also, through electrolysis, provide hydrogen fuel for rockets. Another potential hot spot is the moon's far-side equatorial region, where massive radio telescopes could one day be sited.

In addition to communications, the astronauts, rovers, and scientific instruments all need to know where they are on the moon's surface. Relay satellites can form a sort of "lunar GPS" for navigation by timing how long it takes for signals between multiple satellites to reach a given point on the surface. In general, the more relay satellites in more orbits, the better. The trade-off is that launching and operating each additional satellite costs money. Therefore, any relay-satellite network needs to provide the best possible service and coverage with the minimum number of satellites.

Argotec's relay network concept uses a class of stable orbits known as frozen orbits. Stable orbits make it easy to keep the satellites in their assigned orbits for the 5 years (or more) that they are expected to operate. The proposed orbits are elliptical, with a 12-hour period, a 57-degree inclination, and a distance to the moon's surface from 720 kilometers at their closest points to 8,090 km at their farthest.

Any satellite will travel slowest at the farthest point of its orbit—called the apoapsis—and fastest when it is closest to the moon. Therefore, we want any orbit to have its apoapsis approximately above a potential hot spot in order to provide long periods of communications. With the selected orbits, the lunar poles are covered by three satellites simultaneously 94 percent of the time, with at least one satellite overhead at any given time. The equator, meanwhile, has at least one satellite overhead 89 percent of the time, and simultaneous coverage by three satellites 79 percent of the time.

The proposed FARSIDE telescope would use unspooled antennas across an area of the moon's surface 10 kilometers in diameter to create a large interferometric array.

Even at the apoapsis, a relay satellite is fewer than 10,000 km from the surface. Compare that to the distance from the Earth to the moon, which is about 400,000 km. Even for users positioned within direct line of sight with Earth, an overhead relay satellite reduces the communication link distance by about a factor of 40.

A shorter communication distance means a person or robot on the surface does not need a powerful terminal to maintain a low-data-rate link with Earth. Instead, they can employ the relay satellites to bounce their signals to Earth using a small communications terminal.

Relay satellites also mean that humans at two different locations on the surface can talk to each other without noticeable delay. Without relay satellites, a call would have to travel to Earth and back, taking about 3 seconds round trip. Imagine the difficulty of a phone call with a 3-second delay, and you'll quickly realize how important relay satellites are for voice or video communications on the surface.

Even for users positioned within direct line of sight with Earth, an overhead relay satellite reduces the communication link distance by about a factor of 40.

Different missions will have different communication needs. Simple text or voice communications require only a few kilobits per second, while high-definition video and radio telescopes need megabits per second. And given the number of proposed lunar missions, any relay satellite will likely need to juggle multiple simultaneous communications. For lower bandwidth applications like text and voice, one satellite will be able to collect and aggregate the many data streams for relay elsewhere. On the other hand, an individual satellite is likely to reach its capacity with the high data production of a single radio telescope.

NASA is currently studying two radio-telescope options that could be deployed on the moon's far side. The first is the Lunar Crater Radio Telescope (LCRT), an ultralong-wavelength radio telescope proposed by JPL engineers. The LCRT would observe the universe at frequencies below 30 megahertz, which are otherwise blocked by the Earth's ionosphere. Robots would deploy a wire mesh 1 km in diameter in the middle of a 4-km crater to create a reflector radio telescope. It would be the largest dish-shaped radio telescope in our solar system.

The second proposed telescope is the Farside Array for Radio Science Investigations of the Dark ages and Exoplanets.

FARSIDE would be a low radio frequency interferometric array—meaning it would observe distant stars and other radio sources with multiple antennas. By correlating these multiple observations, it can image the source at high resolution and accurately determine its position. The system would use 128 dual-polarization antennas deployed across a roughly circular area 10 km in diameter, and tethered to a base station for central processing and power. The base station would also transmit collected data to a relay orbiter (such as our proposed Andromeda constellation).

he software-defined Universal Space Transponder radio is the foundation of a lighter and smaller radio called the UST-Lite that JPL is currently testing for use in future spacecraft.

FARSIDE would be able to image the entire sky each minute, spanning frequencies from 100 kilohertz to 40 MHz. Like the LCRT, this would extend into bands below those accessible to Earth-based radio astronomy—in the case of FARSIDE, by two orders of magnitude. Both proposed telescopes would generate massive volumes of data that need to be transmitted to Earth.

After a relay satellite receives data from a far-side radio telescope or anything else on the lunar surface, it will need to send that data onward to Earth. On Earth, large antennas will need to have adequate gain and sensitivity to support a link up to at least 100 megabits per second. Ideally, each (expensive) ground antenna should be able to receive signals from multiple relay satellites at a time to reduce the number that need to be built.

NASA's Deep Space Network (DSN) is a good example of the type of ground network needed. The DSN has three antenna complexes across the world—in California, Australia, and Spain—with several large, highly sensitive antennas at each site. However, the DSN is designed to support deep-space missions well beyond the moon, and so using it for a lunar relay system may be overkill. Besides, the DSN is already in high demand by many missions, both current and planned. So while it may be a good initial choice, over the longer term, leasing or building commercial ground stations would be cheaper and more effective.

A lunar relay spacecraft needs to be only 50 or 60 kilograms, which is small by satellite standards. We have developed a satellite concept that is 44 by 40 by 37 centimeters when the solar arrays and antennas are stowed, with a mass (including propellant) of 55 kg. It carries a four-channel radio developed at JPL, with two channels each operating in the K-band (at about 26 gigahertz) and S-band (at about 2 GHz). One K-band channel provides connectivity to Earth (100 Mb/s for satellite-to-Earth and 30 Mb/s for Earth-to-satellite). The other three channels provide connectivity to the moon. The S-band channels offer 256 kb/s connections to the lunar surface, and 64 kb/s from the surface to the satellite. The remaining K-band channel is a 100 Mb/s satellite-to-moon link and 16 Mb/s moon-to-satellite link.

Imagine the difficulty of a phone call with a 3-second delay, and you'll quickly realize how important relay satellites are for voice or video communications on the surface.

Our proposed satellites would use the K-band for Earth-to-satellite connections for two reasons. First, there is more available bandwidth in the K-band than other bands used for space communications. Second, for antennas of the same size, K-band frequencies have higher antenna gain. In other words, K-band antennas more efficiently convert received signals into electrical power. The downside of using the K-band is its weather sensitivity—rain, for example, will easily attenuate the link. The relay satellites would require an additional power margin to ensure the link remains stable.

The current relay-satellite design has three antennas: A steerable, 50 cm K-band antenna for Earth-to-satellite communications; a fixed K-band " metasurface" antenna that has a low profile with low mass, can easily be manufactured at low cost, and can tolerate the harsh environment of outer space; and a fixed S-band antenna array. We're also considering a small antenna in the X-band (at about 7 GHz) for Earth-to-satellite communications for additional reliability and redundancy. The X-band is a good choice here because it is less susceptible to attenuation from rain than the K-band, albeit at a lower data rate.

Currently, we are finalizing the design of the spacecraft. We intend to use commercially available hardware wherever possible to lower costs. However, we still need a few new technologies to be refined to provide the desired satellite performance while still meeting requirements for mass and power. The metasurface antenna, which can be 3D printed, is a new technology developed at JPL for small-satellite applications. The transmit-only version is operational, with a measured gain exceeding 32 decibels isotropic (dBi) for a 20-cm antenna at 32 GHz. We expect a recent improvement to the design to increase the gain to 34 dBi. We're also working on dual-frequency capability, so that the antenna will be able to simultaneously transmit and receive signals.

The three antenna complexes making up the Deep Space Network, such as the one in Canberra, Australia that includes the antenna shown here, maintain contact with spacecraft across the solar system. The Andromeda constellation would need a similar setup to bring back data from the moon.

Additionally, we'd like to use a smaller and lightweight version of the software-defined Universal Space Transponder (UST) radio called UST-Lite. JPL has completed an initial thermal-testing campaign for a UST-Lite prototype, to ensure that the radio's generated heat can be dissipated without affecting performance. We performed additional tests to better characterize the prototype's receiver thresholds, bit error rates, transmit waveforms, and more. We continue to optimize the receiver's parameters, as well as to develop new modules to cover K-band frequencies (We have already developed S- and X-band modules).

We're also addressing the network's software needs. For example, there is no current protocol standard for communications between a relay satellite and a lunar user at the S- and K-bands. We, therefore, have begun to work with the Consultative Committee for Space Data Systems to introduce such a standard.

One way to think about the goal of any lunar-communications apparatus is that it would create 5G-like capabilities for the entire moon. This would mean taking advantage of 5G technologies wherever possible, such as installing cell sites on the moon to supplement the relay arrangement. This approach would connect many additional kinds of devices to a lunar network—for example, networks of low-power Internet of Things sensors and autonomous vehicles.

Our proposed relay network would only be a first step. In a more distant future, humans on the moon should be able to send and receive texts, make phone calls, and stream data at will. Similarly, robots and sensors should be wirelessly connected just like IoT devices are on Earth. Robots would be controlled remotely, and sensors would automatically upload their measured data.

However, this vision of lunar connectivity may take generations of lunar-communication networks to emerge. Nevertheless, we believe we can look forward to a time when there will be human colonies on the moon engaged in scientific, technical, and commercial activities in a robust wireless environment.

14.0 Mining Water Ice

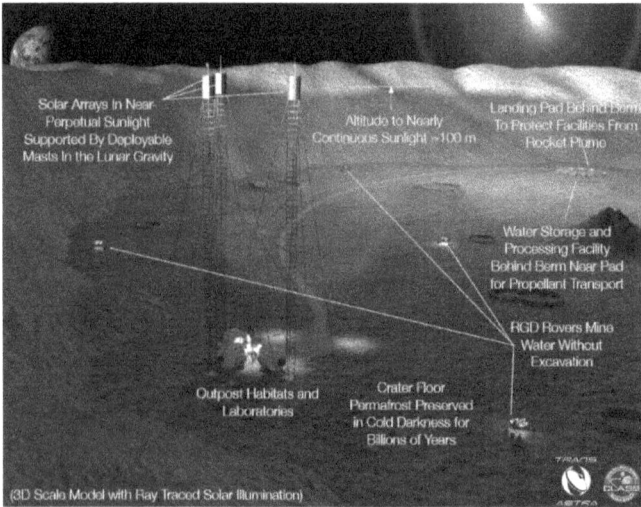

The Lunar Polar Gas-Dynamic Mining Outpost (LGMO) is a breakthrough mission architecture that promises to greatly reduce the cost of human exploration and industrialization of the Moon. LGMO is based on two new innovations that together solve the problem of affordable lunar polar ice mining for propellant production.

The first innovation is based on a new insight into lunar topography: the analysis suggests that there are large (hundreds of meters) landing areas in small (0.5-1.5 km) near polar craters on which the surface is permafrost in perpetual darkness but with perpetual sunlight available at altitudes of only 10s to 100s of meters. In these prospective landing sites, deployable solar arrays held vertically on masts 100 meters or so in length (lightweight and feasible in lunar gravity) can provide nearly continuous power. This means that a large lander, such as the Blue Moon vehicle proposed by Blue Origin, a BFR; or a modestly sized lunar ice mining outpost could sit on mineable permafrost with solar arrays in perpetual sunlight on masts providing affordable electric power without the need to separate power supply from the load.

The second enabling innovation for LGMO is Radiant Gas Dynamic (RGD) mining. RGD mining is a new Patent Pending technology invented by TransAstra to solve the problem of economically and reliably prospecting and extracting large quantities (1,000s of tons per year) of volatile materials from lunar regolith using landed packages of just a few tons each. To obviate the problems of mechanical digging and excavation, RGD mining uses a combination of radio frequency, microwave, and infrared radiation to heat permafrost and other types of ice deposits with a depth-controlled heating profile. This sublimates the ice and encourages a significant fraction of the volatiles to migrate upward out of the regolith into cryotraps where it can be stored in liquid form. RGD mining technology is integrated into long duration electric powered rovers.

In use, the vehicles stop at mining locations and lower their collection domes to gather available water from an area before moving on. When on-board storage tanks are full, the vehicles return to base to empty tanks before moving back out into the field to continue harvesting. The rover can be battery operated and recharge at base or carry a laser receiver powered by a remote laser. Based on these innovations, LGMO promises to vastly reduce the cost of establishing and maintaining a sizable lunar polar outpost that can serve first as a field station for NASA astronauts exploring the Moon, and then as the beachhead for American lunar industrialization, starting with fulfilling commercial plans for a lunar hotel for tourists.

Tools and Equipment

The tools necessary to carry out any significant production of raw materials for manufacturing cannot be regarded as simple, but merely as relatively simple.

One item will have to be dirt-moving equipment, necessary for habitat excavation, as well as for transporting feedstocks to smelting or manufacturing sites and removing waste products and, perhaps, the products themselves. It is assumed that all processes, from gathering and production of raw materials through manufacture and assembly, will be substantially automated. Equipment probably will be tended, mainly from remote stations nearby.

Equipment should be designed to allow straightforward repair, optimization to actual encountered conditions, and innovative adaptation by the operator to new feedstocks and conditions. Electrical power, at least one megawatt, will have to be available for any significant processing or manufacturing activity. Initially, this probably will need to be nuclear power, although eventually solar power should be exploited to the fullest extent. Through the use of concentrating mirrors, solar power should be available at the outset for heating of materials to high enough temperatures to melt or even distill them.

15.0 Growing Food on the Moon

Below is an interview about growing food for the moon.

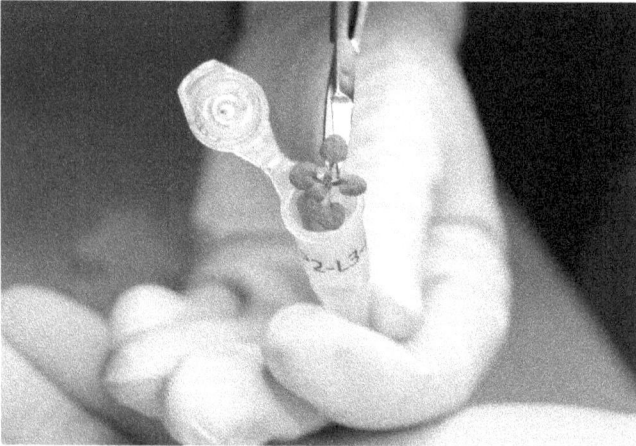

Researchers at the University of Florida successfully grew plants in lunar regolith brought back during three different Apollo missions. In this photo, a scientist places a plant grown during the experiment in a vial for eventual genetic analysis.

Space botanists are working on strategies to grow crops on the lunar surface, as NASA makes strides toward sending astronauts to the Moon through the Artemis program. A team of scientists at the

University of Florida successfully grew small plants in lunar soil brought back during three different Apollo missions. How did they do it, and what does it mean for the future of space exploration? Dr. Anna-Lisa Paul explains.

Jim Green: Can we grow food on the Moon? This may end up being a fundamental question of survival in space. Let's talk to a space botanist.

Anna-Lisa Paul The only way that humans can be explorers is if we take our plants with us.

Jim Green: Hi, I'm Jim Green, and this Gravity Assist, NASA's interplanetary talk show. We're going to explore the inside workings of NASA and meet fascinating people who make space missions happen.

Jim Green: I'm here with Dr. Anna-Lisa Paul. And she is the professor of horticultural sciences at the University of Florida's Institute for Food and Agricultural Sciences. And she is the director of the University of Florida's Interdisciplinary Center for Biotechnological Research.

Jim Green: Dr. Paul and her colleagues just published a fantastic new study. And this study describes how plants grow in samples of lunar soil brought back by astronauts in the Apollo program. Wow! I can't wait to hear how this was pulled off. So welcome Anna-Lisa to Gravity Assist.

Anna-Lisa Paul: Thank you. Thank you very much. Pleasure to be here.

Jim Green: The paper that's out now is really exciting, because it tells us that we now have options of going to the Moon and being able to live and work on a planetary surface for long periods of time, because we have an aspect of sustainability by growing food.

So is this project something you've been wanting to do for a long time?

Anna-Lisa Paul: Absolutely. This is a project that has been sort of on my, and my colleague, Rob Ferl's radar, for decades, because when you think about if the only way that we can humans can be explorers, is if we take our plants with us. Plants are what allows us to be explorers, they can go past the limits of a picnic basket. So for us who work in space biology, we wanted to know if when we get to a new surface, can we use the resources that are already existing there, the in situ resources? And for the Moon, that would be the regolith, which can be used as the dirt to grow plants.

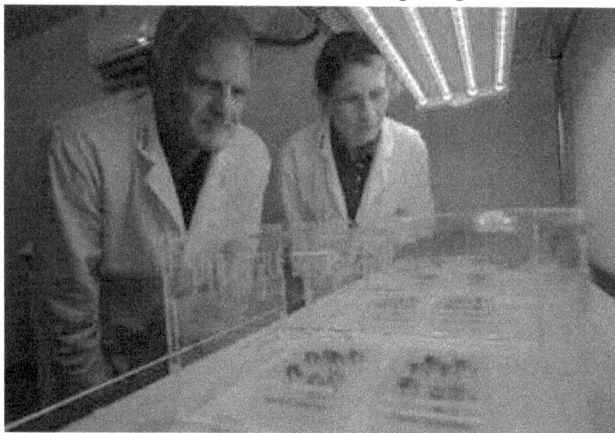

University of Florida researchers Rob Ferl, left, and Anna-Lisa Paul, examine a collection of culture plates – some filled with lunar regolith, some with simulated regolith -- under LED lights.

Jim Green: Well, how hard was it to get your hands on these samples, the original samples from the Apollo program?

Anna-Lisa Paul: It was pretty hard to get those. You have to remember, they're a national treasure, they are completely irreplaceable in their original form. And so when you have a couple of biologists who go to an institution of higher archiving from NASA of the original Apollo samples, and you say, "Yes, we'd please like to have some of your precious materials and get them all messy and grow plants in them!" They say, "Excuse me, you want to do what?" And so it took three different iterations of proposals, which also include a ton of background information and tests with lunar simulants before we could convince the powers that be that, yes, yes, we will take good care of them. We're good representatives of what science can be done, and they let us have some. In fact, they let us have 12 grams.

Jim Green: 12 grams. I know that doesn't sound a lot.

Jim Green: Well, what's really amazing to me when we think about plants growing in regolith is, is what regolith is. You know, it's really ground up rock, that comes from impacts over and over, billions of years of impacts on the Moon, blasting everything apart. And when you look at the regolith, this ground-up rock, in a microscope, it's got all these shards. It's, it's very sharp, which is one of the reasons why we're worried about this regolith, when humans walk around in spacesuits, getting into their lungs.

Jim Green: And so the concept that we can actually grow plants in it, was really amazing. So, tell us about these lunar samples. Did they come from one location or many locations?

Anna-Lisa Paul: So the samples actually came from three locations: from Apollo 11, Apollo 12, and Apollo 17. And so the three sites that the astronauts worked on had different characteristics. All of the materials are what are called basaltic. And so most of them were sort of ground up basalt, lava kind of, kind of materials. But each of the sites were exposed to the surface for different periods of time. And what that means is that the regolith has what's called different levels of maturity.

Anna-Lisa Paul And so the regolith from the Apollo 11 site, for instance, was more mature. That means it has been exposed to the cosmic wind for longer. So the particles are smaller, the edges are sharper. The Apollo 17 samples were particularly interesting in that it, the type we got was actually a compendium of materials from all over the site, because it was the dirt, if you will, that got caught underneath a bumper on the lunar rover.

Anna-Lisa Paul: And as, as they were leaving, Harrison Schmidt said, wow, there's a whole bunch of stuff here. Let's not let that go to waste. And he dumped it all into a bag and it came back to Earth for, for us eventually.

Jim Green: Wow, that's fantastic. So tell me about the experiment. If you only had a little bit from each of these sites, how are you going to really grow plants in them?

Anna-Lisa Paul: So we used the plant called Arabidopsis thaliana. And the cool thing about Arabidopsis is, in addition to being very well characterized at the genomic level, and gene level, it's small, it's really small, and you can actually grow an almost full size plant in a single gram of material.

Jim Green: Wow.

Anna-Lisa Paul: So what we did is we had these specialized plates that are normally used for cell culture, there are only about 12 millimeters across -- each one of these little pots, if you will. And we put the regolith inside these little pots and then planted seeds on top of them, watered them from below and: instant lunar garden.

Jim Green: Wow, that's unbelievable. So you had a regimen of just adding water to the to the seed and that's all it took?

Anna-Lisa Paul: It took a little bit of nutrients, too.

Jim Green: Okay.

Anna-Lisa Paul: And so how it was set up was a little plug of material called rockwool, which is essentially just spun lava rocks, that makes a sponge, and then the regolith goes on top of that little sponge. And so now the sponge acts as a capillary wick to get liquids up into the regolith. So the nutrient solution that went down into the base of the tray got wicked up into the regolith, and it was essentially watered from below.

Jim Green: Wow, interesting. So then it's easy to think about how that could work by developing a greenhouse with these kind of attributes on the Moon and then just bringing in the regolith.

Jim Green: So at the end of the experiment, did you then take apart the regolith to see how the roots grew with in the planter?

Anna-Lisa Paul: We did. Because we planted more than just a single seed at first, when we thinned the little tiny seedlings away to just leave a single plant in each one of those little micro pots, we also got to look at the roots there. And so we could see that the plants that were growing in the simulant, it's called this JSC-1A, it's a type of volcanic ash that's mined on Earth, that's what we use as our control.

Anna-Lisa Paul: Compared to the lunar regolith, the JSC-1 simulants were nice and long and tapered and looked very healthy, but the roots that were growing in the regolith were kind of scrunched up and they weren't quite as healthy looking. Nonetheless, once they grew, you could get decent looking plants growing in the regolith. And just to look at them with your eye, they'd look a little smaller than the ones in the controls. But the real key was when you ground them up, and you look at what genes are being expressed.

Jim Green: Now, as you said, you use simulant, which means we think we've been able to develop a process that can make lunar-like regolith without bringing it from the Moon. But as you said, already, there's some differences between that simulant and what the real regolith looks like. But that's an important control factor. That also helps us figure out if we're making those simulants correctly or not.

Anna-Lisa Paul: Yup.

Jim Green: So what did you find out?

Anna-Lisa Paul: So when you take a look at the controls, I have to say, any experiment is only as good as your control, right?

Jim Green: Right.

University of Florida researchers Anna-Lisa Paul and Rob Ferl are seen at the Haughton Crater impact site in northern Canada. NASA uses this crater for Moon and Mars analog research.

Credits: Pascal Lee

Anna-Lisa Paul: And so, the control material really did look a lot like the lunar regolith. It behaved a lot like the lunar regolith in the way it absorbed water and the way that it kind of just settled into the pots and everything. But when we've looked at the example of even if you take two plants that looked very similar between the control and the lunar regolith grown, we found that the kind of genes that the plants expressed different from the ones that were in the control were mostly genes that are associated with metal stress, like heavy metals, or salts, or what we call oxidative stress.

Jim Green: Oooh!

Anna-Lisa Paul: Even though those materials per se weren't necessarily in those regoliths. It's not like the regoliths were actually salty. But the plants perceived the type of stress they were seeing in that material as salt stress, as metal stress. And so that was an interesting insight that they were changing the way they express their genes to adapt to that new and novel environment.

Jim Green: Oooh. So this is really critical to understand. Because once you understand that, there may be processes and procedures that you could do that alleviate that plant stress that allows them on, on the real example, on the Moon in a greenhouse, to then really flourish better than even what you did in the laboratory.

Anna-Lisa Paul: That's exactly right. That's button on. So the Arabidopsis is really closely related to some of your favorite vegetables, like, say, broccoli. And we know that if we want our broccoli plants or kale plants to be healthy and growing in the lunar regolith, in a greenhouse, we know that we'll have to mitigate some of these kind of stress responses. We can do that in two ways. You can engineer their environment by mitigating perhaps some of the materials that are in the regolith, you can also engineer the plants themselves. And you can make them less sensitive to some of these aspects. And so instead of putting their energy into the stress response, they put that energy into making more broccoli.

Jim Green: Right! That's really a, just a huge advance. By doing this on the Moon, we're going to also learn the processes and procedures we'll have to do on Mars. So that will be really critical. S o I really dearly love this idea. So if I was in the lab, and we were done with the experiment, we were taking them apart and looking at the roots, I might be tempted to eat one of these. Did anyone do that?

Anna-Lisa Paul: Well, we didn't eat any of those because, think about it: they're a very small and very precious resource that we wanted to save to do the biochemical analyses. You could eat Arabidopsis. People have eaten them before, but it's not exactly something that would be good in a salad.

Jim Green: (laughs) So not so tasty after all.

Jim Green: I can imagine walking into the lab, when it, when you had started these plants growing. And the first time you realized this was gonna work. What was that like?

Anna-Lisa Paul: Oh, so the preparation that went into this experiment is extraordinary. All the background, all the setup, everything, the way we planted them, every aspect of it was complex. And so then at the end, Rob, and I walk out to our secure growth chamber where these things are going to go, we set them all up under their pink LED lighting systems that will keep them going. And we closed the door and we thought, all right, three days, things should be germinating in three days. Well, two days later, we walked back in there just to kind of check, and we're looking down at all those plates. And every single one had germinating seeds in it.

Jim Green: Wow!

Anna-Lisa Paul: The controls, the lunar samples, everything was germinating. There's this tiny nascent greenness, every single one, and it just took our breath away. It worked. It really worked. How cool is that?

Jim Green: You know, it reminds me of the theme in the movie "The Martian," where Mark Watney goes over to his potato plant that is now growing for the very first time, touches the leaf, and says "hello."

Anna-Lisa Paul: Yes, exactly.

Jim Green: Wow, that's great. I can also imagine that this will enable you to think of the next best experiment to do. Have you been thinking about and formulating your next steps?

Anna-Lisa Paul: Oh, absolutely. One of the things that would be wonderful to do is to have additional replicates for this. With four grams each from each site, we could obviously only have four replicates of one individual plant each. Being able to have a larger volume of material so that we could try different kinds of mitigations. All of the samples had to be treated with the same nutrient solution for instance. And so if we had enough material, we could also change the variables of what kind of nutrients we did. Are there other ways to mitigate some of the effects of the regolith? Those are the kinds of things you can only do with more material.

Jim Green: I understand you've done some field tests in far off places here on Earth.

Anna-Lisa Paul: Yeah, so I've definitely had the privilege to explore some very interesting, what we call analog sites, in the in the world. The first step was, Rob Ferl and I went to the far north Canadian Arctic at an old impact site, called the Haughton Crater on Devon Island. And one of the reasons we went to Devon Island was to practice utilizing in situ resources in a greenhouse that was growing there.

Anna-Lisa Paul: And so we collected these, what we call, brecciated materials from this old impact crater, which was 20-plus miles across, that was very lunar looking. And we've use some of those materials in the greenhouse. We also used the JSC-1 simulant in the greenhouse, along with other kinds of materials and asked: Can we populate a greenhouse substrate with these kinds of non-traditional growth substrates to create materials and crops over the winter?

Jim Green: So what did you find out when you did that?

Anna-Lisa Paul: Well, we find that they actually like growing in the JSC-1 simulant a little better than they liked growing in the brecciated materials we dug out of the crater. (laughs) And part of that is because a lot of the materials have different types of chemicals in them that are actually in some ways more analogous to what it would be on Mars. Whereas the lunar regolith is pretty much just devoid of everything, the Martian regolith i, looks to be, although nobody's brought any back, it looks to be high in, say, perchlorates and other kinds of reactive chemicals that would have to be, again, ameliorated before you could grow plants in it. But you'd be have to be able to use the materials from where you land.

Jim Green: So on the Moon, I imagine we're going to have a greenhouse, but can we really grow these out in the vacuum of space?

Anna-Lisa Paul: Well, they would have to have a greenhouse just like a human would have to have a greenhouse because that there's no atmosphere on the surface of the Moon. So all of the plant growth would be being carried on in some kind of greenhouse or other sort of enclosed habitat along with its attending humans.

Jim Green: Well, you know, another part about that, that I like, is the fact that these plants as they grow will smell wonderful. And you get not only this the green of the plant, you also get the smells, and it's gotta remind astronauts of home.

Anna-Lisa Paul: That that is so true. And I have actually a personal experience that, that speaks to that very well. I mentioned the work that I've done in the high Canadian Arctic. Well, I've also been down in Antarctica for a while. And again, working on a greenhouse that was essentially called the Future Exploration Greenhouse, part of the Eden ISS project, that was an analogue of what you might find on the Moon or Mars.

Anna-Lisa Paul: I was down there for several days, and the weather was just horrible, and nobody could go outside, it was absolutely impossible, and everything was dark, and bleak and awful. And then, when the weather started to clear just a little bit, we went out to the greenhouse for the first time on that trip and walked into the door, and you're met by the smells and the moisture and the greenness. And it was like, all of the stress evaporated from all of us. And we were home for a bit. And I can well imagine it would be like that for an astronaut. And you can't underestimate how powerful, how powerful a plant can be from that context, as well as the fact that it cleans your air and gives you clean water and gives you food. It also gives you something spiritual.

Jim Green: Very nice.

Jim Green: Well, Anna-Lisa, I always like to ask my guests to tell me what that person place or event was that got them so excited about being in the sciences that they are today. And I call that event, a gravity assist. So Anna-Lisa, what was your gravity assist?

Anna-Lisa Paul: Well, gravity assist for me has been people, and the very first person was my mom. And I can remember quite keenly as a little kid asking my mother about how something worked. And she would say, "I don't know, let's find out." And so it was always this, this journey of discovery. I would be given science books as a small kid, even though I couldn't quite read them at that level. And we'd go through as a family trying to figure out how to do the kind of experiments we could do in the backyard. And I got really interested in plants, because plants were the only things that were taking the energy that comes into the planet, and turning it into stuff that we needed.

Anna-Lisa Paul: So as I got older and started wondering about how plants work, it kept taking me one step after another until I decided I'd like to understand how plants respond to novel environments, and the most novel environment out there is space.

Jim Green: Wow, fantastic. That, that's a wonderful environment to be in, where you can work with your parents on a journey of discovery, and then realize how you can make a wonderful career out of it. So thanks so much for telling us about this really fundamental and exciting research.

Anna-Lisa Paul: I'm pretty lucky. Thanks.

Jim Green: You're very, very welcome. Well, next time, we're going to talk to a researcher at the Kennedy Space Center, who also works on growing plants in space. But in this case, it's all about astronauts growing them on the space station. You won't want to miss that. I'm Jim Green, and this is your Gravity Assist.

16.0 Current Plans for NASA Moon Bases

NASA's plans for a Moon base:

NASA's official plans to build a permanent base on the Moon have leaked online, revealing how and when astronauts will return to the rocky world for the first time in 50 years.

Internal documents show how NASA wants to launch 37 rockets to the Moon within the next decade, with at least five of these carrying astronauts.

Starting with an unmanned rover in 2023, the space agency is expected to land people on the Moon in 2024.

NASA will then fire manned missions to Earth's neighbor every year between 2024 and 2028, according to the documents, which were obtained by Arstechnica.

Speaking to The Sun, a NASA spokeswoman confirmed the documents are real and revealed the plans were briefed today during a public session of the Science Committee to the Nasa Advisory Council (NAC).

They show a decade-long program that culminates with a permanent lunar base, which NASA will begin building in 2028.

They are in part a response to recent calls from U.S. Vice President Mike Pence to take astronauts back to the Moon.

"In the nearly two months since Pence directed Nasa to return to the Moon by 2024, space agency engineers have been working to put together a plan that leverages existing technology, large projects nearing completion, and commercial rockets to bring this about," Arstechnica's Eric Berger wrote.

"Last week, an updated plan that demonstrated a human landing in 2025, annual sorties to the lunar surface thereafter, and the beginning of a Moon base by 2028, began circulating within the agency."

Berger did not say how he obtained the plans, which have not yet been made public.

They do appear to line up with previous statements from NASA about its lunar program, codenamed Artemis.

As with any space exploration project, the main obstacle is cash.

NASA reckons it will need $4.7 billion to $8.2 billion per year on top of NASA's existing budget of about $20 billion.

Boss Jim Bridenstine recently asked for an extra $1.6 billion in fiscal year 2020 to start developing a lunar lander.

The plan also relies heavily on contractors delivering ambitious hardware on time, which has hindered NASA in the past.

Boeing has been developing the core stage of the agency's next-gen rocket, the Space Launch System, for eight years – but has yet to come up with the goods.

ARTEMIS AFTER 2025

After Artemis III, the overall plan is to conduct operations on and around the Moon that help prepare us for the mission durations and activities that we will experience **during the first human mission to Mars, while also emplacing and building the** infrastructure, systems, and robotic missions that can enable a sustained lunar surface presence. To do this, we will develop Artemis Base Camp at the South Pole of the Moon.

Artemis Base Camp will be our first sustainable foothold on the lunar frontier. We will initially move to one to two-month stays to learn more about the Moon and the universe. We will develop new technologies that advance our national industries and discover new resources that will help grow our economy. Overall, the base camp will demonstrate America's continued leadership in space and prepare us to undertake humanity's first mission to Mars.

A South Pole landing site has not been determined, but this image shows sites of interest near permanently shadowed regions, which may

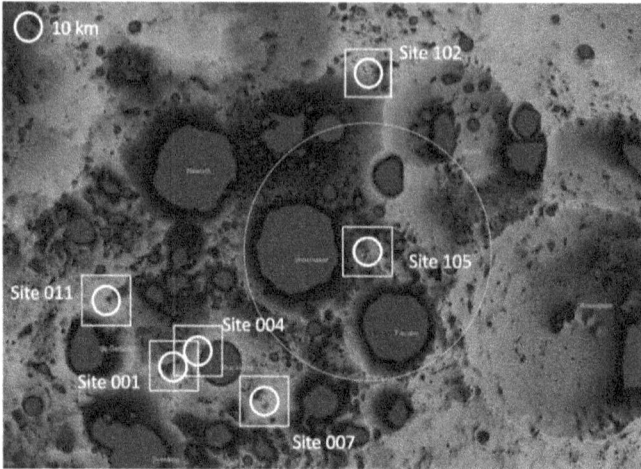

contain mission-enhancing volatiles. These sites may also offer long-duration access to sunlight, direct-to-Earth communication, surface slope and roughness that will be less challenging for landers and astronauts.

The three primary mission elements of Artemis Base Camp are: The LTV that can transport crew around the site; the habitable mobility platform for long-duration trips away from Artemis **Base Camp and the foundation surface habitat will enable short-stays for four crew on the lunar South Pole. Combined with supporting infrastructure added over time such as communications, power, radiation shielding, a landing pad, waste disposal, and storage planning – these elements comprise a sustained capability on the Moon that can be revisited and built upon over the coming decades.**

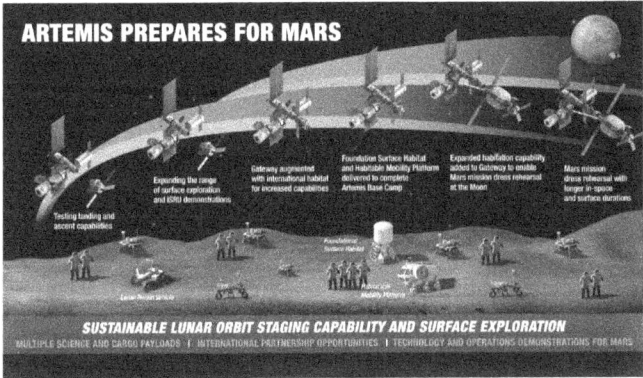

ARTEMIS PREPARES FOR MARS

SUSTAINABLE LUNAR ORBIT STAGING CAPABILITY AND SURFACE EXPLORATION

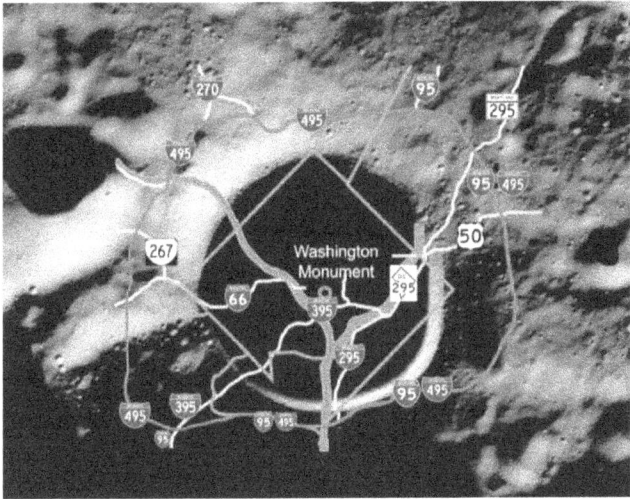

The lunar South Pole's Shackleton Crater, as captured by the Lunar Reconnaissance Orbiter, with the Capital Beltway overlaid for scale.

Mobility is a major part of the Artemis Base Camp. The LTV and the habitable mobility platform will enable long-term exploration and development of the Moon. In addition to its size, the Moon's geography is complex, and its resources dispersed. Looking at potential sites for Artemis Base Camp, such as near Shackleton Crater, shows the immense scale of the lunar geography. Robust mobility systems will be needed to explore and develop the Moon. The same is true for Mars, making the habitable mobility platform a particularly important element as we will need a similar type of vehicle to explore the Red Planet.

In addition to establishing Artemis Base Camp, another core element of the sustained lunar presence that feeds forward to Mars will be the expansion of habitation and related support systems at the Gateway. This evolution of the Gateway's systems to include large-volume deep space habitation would allow our astronauts to test, initially in lunar orbit, how they will live on their voyage to and from Mars. Gateway can also support our first Mars mission analogs on the lunar surface. For such a mission, we currently envision a four-person crew traveling to the Gateway and living aboard the outpost for a multi-month stay to simulate the outbound trip to Mars, followed by two crew travelling down to and exploring the lunar surface with the habitable mobility platform, while the remaining two crew stay aboard. The four crew are then reunited at the Gateway for another multi-month stay, simulating the return trip to Earth, before landing back home.

These missions will be by far the longest duration human deep space missions in history. They will be the first operational tests of the readiness of our long-duration deep space systems, and of the split crew operations that are vital to our approach for the first human Mars mission.

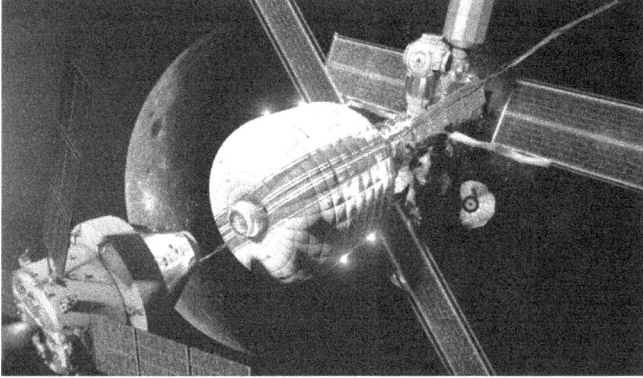

Figure 8: Orion approaches an evolved Gateway.

There are many factors associated with the sequence of element development, testing, and launch such as capability maturity and availability, budget, launch vehicle availability, and system complexity. For planning purposes, NASA is developing a sequence that accounts for these variables and results in an annual cadence of demonstrable progress and a gradual increase in mission duration and complexity. This plan results in the development and emplacement of the infrastructure required for a long-term sustained lunar surface presence while testing systems and gaining the operational experience required for the human Mars mission.

The sequence as currently envisioned begins by sending lunar precursor robotic missions including VIPER by CLPS landers to provide ground truth of terrain, as well as water and metal resource availability for the human lunar landing site. To provide mobility and extended range of exploration for the first several human lunar surface missions, the LTV will be delivered to the lunar surface. The first elements of the lunar Gateway are in development and will support later sustainable human lunar landing missions. NASA anticipates its international partners will provide at a minimum the robotic arm, I-Hab, and ESPRIT to supplement the Gateway's capabilities in lunar orbit.

The habitable mobility platform will be delivered to the lunar surface to expand our exploration range by tens of kilometers and mission duration on the surface from 7 days to 30-45 days, enabling potential Mars surface analog missions on the lunar surface. Other key pieces of the Artemis Base Camp infrastructure are also delivered, including the foundation surface habitat, which will support a crew up to four on the lunar surface, the lunar surface power systems, ISRU demonstrations and pilot plants.

An evolved Gateway habitation capability in lunar orbit will allow us to begin the methodical lengthening of mission durations. This approach will also allow NASA to test risk mitigation approaches for long-duration mission crew and element systems risks that are required for two-year Mars class missions.

Once these pieces of the Moon to Mars campaign are delivered and operational, annual human missions with increasingly long durations will enhance the exploration and sustainable development of the lunar surface.

A VIBRANT EARTH-MOON FUTURE

Whenever the first human mission to Mars occurs, it will not mean that we are done with the Moon. The windows for launching the two-year mission to Mars open up every few years, and we will continue to conduct human missions to the lunar surface to test systems, conduct scientific investigations, and continue to develop our sustainable lunar presence as we prepare for the optimal launch window.

We will continue to explore the Moon indefinitely -- leveraging robotic deliveries provided by CLPS providers, longer duration human missions, and commercial and international.

17.0 Chinese and Russian Lunar Plans

NASA and it's partners are not alone in initiating new exploration and settlement of the Moon. The Chinese and Russians are also active in planning a Moon Base within the next decade.

In this chapter are some interesting finds of one of the Chinese unmanned landers on the far side of the moon and more details about the planned Chinese and Russian moon base.

17.1 Unmanned Lunar Lander Visits to the Farside

Picture of Gel on the Far side of the Moon

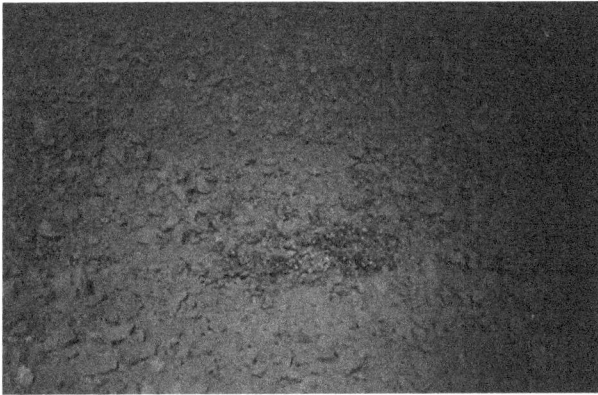

In July of 2019 China's Yutu-2 rover discovered something with an unexpected color and luster during its travels on the far side of the Moon. On September 1, a tweet from People's Daily – largest newspaper group in China – used the words "gel-like" to describe this substance. Weird! The choice of words piqued a lot of curiosity, although some scientists stated at the time the rover had probably stumbled on something more like impact glass, created after a meteorite hits the lunar surface.

Now, it appears those scientists were right. The China Lunar Exploration Program has released a new photo of the substance, and the bright specks do resemble other impact glass – known as impactite and resembling trinitite on Earth – that's been seen on the Moon before. The photo, taken by the Yutu-2 rover's main camera, shows the center of the small crater, with numerous small bright spots on the lunar regolith.

The image doesn't look too unusual, just showing the grey regolith with the small bright flecks in the center of the crater. It was analyzed and processed to bring out more detail by Daniel Moriarty, a NASA Postdoctoral Program fellow at the Goddard Space Flight Center. As he explained:

The shape of the fragments appears fairly similar to other materials in the area. What this tells us is that this material has a similar history as the surrounding material. It was broken up and fractured by impacts on the lunar surface, just like the surrounding soil.

I think the most reliable information here is that the material is relatively dark. It appears to have brighter material embedded within the larger, darker regions, although there is a chance that is light glinting off a smooth surface. But we're definitely looking at a rock.

17.2 A Chinese and Russian Lunar Base

Russia, China reveal moon base roadmap but no plans for astronaut trips yet

There are no plans to launch astronauts anytime soon.

China and Russia will start preparation work for their future lunar research station this year

China and Russia have invited international partners to join them in building a moon base but revealed they don't plan to send astronauts to the moon in the next decade.

The International Lunar Research Station (ILRS) will consist of a space station in lunar orbit, a moon base on the surface and a set of mobile rovers and intelligent "hopping" robots, according to representatives of Russia's space agency Roscosmos and China National Space Administration (CNSA).

Speaking at the Global Space Exploration Conference (GLEX) in St. Petersburg, Russia, on Wednesday (June 16), Chinese and Russian space officials said they were already in negotiations with international partners including the European Space Agency (ESA), Thailand, the United Arab Emirates and Saudi Arabia to join their endeavor.

The two space powers signed an intergovernmental memorandum of understanding in March 2021 to go ahead with the project.

The timeline presented at the GLEX forum foresees a reconnaissance phase to begin in 2021. By 2025, the space agencies will choose a site for the moon base, with construction expected to follow between 2026 to 2035. The ILRS will become operational from 2036 onwards, providing a range of scientific facilities and equipment to study lunar topography, geomorphology, chemistry, geology and internal structure of the moon, as well as enabling space and Earth observations from the moon's surface. It will also likely support human exploration in the future.

The base on the surface will be serviced via an orbiting station in cislunar space between the moon and Earth, that will see regular traffic between the two celestial bodies.

CNSA vice administrator Yanhua Wu said at the conference that the partners are currently focusing on developing robotic lunar exploration technology and don't plan to send astronauts to the moon within the next decade.

"We will also do a lot of preparatory work and research work in this aspect," Yanhua said. "So we hope to be able to actually send our researchers to the surface of the moon in the future for them to carry out missions on the surface of the moon."

Sergey Saviliev, deputy director general for international cooperation at Roscosmos said at the conference that while China and Russia have not yet been approached by any private companies, the nature of the ISLR project is meant to be inclusive and open to everyone interested.

Earlier this year, Russia announced plans to build its own space station in low Earth orbit, the region of space below 620 miles (1000 kilometers) and threatened to leave the International Space Station cooperation as a retaliation against U.S. sanctions.

18.0 Additional Technologies Needed

Many technologies which were already developed and prototyped on the International Space Station can also be used on a Moon Base:

Heat and Air Conditioning

The ISS has a lot of design elements used to maintain and control temperature. Without thermal controls, the temperature of the orbiting Space Station's Sun-facing side would soar to 250 degrees F (121 C), while thermometers on the dark side would plunge to minus 250 degrees F (-157 C). There might be a comfortable spot somewhere in the middle of the Station, but searching for it wouldn't be much fun!

Fortunately for the crew and all the Station's hardware, the ISS is designed and built with thermal balance in mind -- and it is equipped with a thermal control system that keeps the astronauts in their orbiting home cool and comfortable. The first design consideration for thermal control is insulation -- to keep heat in for warmth and to keep it out for cooling.

Here on Earth, environmental heat is transferred in the air primarily by conduction (collisions between individual air molecules) and convection (the circulation or bulk motion of air). "This is why you can insulate your house basically using the air trapped inside your insulation," said Andrew Hong, an engineer and thermal control specialist at NASA's Johnson Space Center. "Air is a poor conductor of heat, and the fibers of home insulation that hold the air still minimize convection."

"In space there is no air for conduction or convection," he added. Space is a radiation-dominated environment. Objects heat up by absorbing sunlight and they cool off by emitting infrared energy, a form of radiation which is invisible to the human eye.

As a result, insulation for the International Space Station doesn't look like the fluffy mat of pink fibers you often find in Earth homes. The Station's insulation is instead a highly-reflective blanket called Multi-Layer Insulation (or MLI) made of Mylar and Dacron.

The reflective silver mesh is aluminized Mylar. The copper-colored material is kapton, a heavier layer that protects the sheets of fragile Mylar, which are usually only 0.3 mil or 3/10000 of an inch thick.

"The Mylar is aluminized so that solar thermal radiation can't get through it," explains Hong. Here on Earth, we use blankets containing aluminized Mylar to wrap people who have been exposed to cold or trauma. Such blankets are especially popular among hunters and campers!

"Layers of Dacron fabric keep the Mylar sheets separated, which prevents heat from being conducted between layers," he continued. "This ensures radiation will be the most dominant heat transfer method through the blanket." Except for its windows, most of the ISS is covered with the radiation-stopping MLI.

Water Purification

All water used to be hauled into space by rocket then used up or wasted. The ISS now has a water production system in usage since 2010.

Drinkable water is one of the primary and most important assets for human survival. So when preparing for a journey, whether to sea or to space, planners must take this vital resource into consideration. Stowage space during such voyages always comes at a premium. It is no different for the International Space Station and the resupply vehicles that dock there.

A great example of a solution to minimize size and weight in life support is the recently launched Sabatier system. Originally developed by Nobel Prize-winning French chemist Paul Sabatier in the early 1900s, this process uses a catalyst that reacts with carbon dioxide and hydrogen - both byproducts of current life-support systems onboard the space station - to produce water and methane. This interaction closes the loop in the oxygen and water regeneration cycle. In other words, it provides a way to produce water without the need to transport it from Earth.

The fundamental technology for this particular system has been in development for the past twenty years. The overall schedule for hardware production, however, was under two years. This accelerated timeline was a significant challenge for the complex Sabatier, which contains a furnace, a multistage compressor, and a condenser/phase-separation system. The fact that recycling system feeds for Sabatier were already available on the station helped to simplify some of the design tasks by reducing the unknowns.

According to Jason Crusan, chief technologist for space operations at NASA Headquarters in Washington, the previous development and solid interfaces allowed NASA to try out a new way of acquiring services for the station with Sabatier. "Being able to demonstrate innovative new methods to acquire technical capabilities is one of the key cornerstones the space station can serve for future missions and approaches to those missions," Crusan explained.

Using developing technologies and productive systems enables the station to squeeze every drop from the resources that must launch from Earth. In addition to improving the efficiency of the station's resupply capabilities, Sabatier also frees up storage space. This helps to maximize the area available for science facilities and engineering equipment. The knowledge gained from such systems also advances the collective understanding of technologies to advance spaceflight and help solve similar problems on Earth.

The Sabatier system has long been a part of the space station plan, but the retirement of NASA's space shuttles elevated the need for new resources to provide water. For a decade, shuttles have provided water for the station as a byproduct of the fuel cells they use to generate electricity. Sabatier supplements the capability of resupply vehicles to provide water to the station, without becoming a sole source for this critical station resource.

Currently in operation on the station, Sabatier is the final piece of the regenerative environmental control and life-support system. This hardware was successfully activated in October 2010 and interacts directly with the Oxygen Generation System, which provides hydrogen, sharing a vent line.

Prior to Sabatier, the Oxygen Generation System vented excess carbon dioxide and hydrogen overboard. Rather than wasting these valuable chemicals, Sabatier enables their reuse to generate additional water for the station. With room and resources at a premium in space, this is a significant contribution to the space station's supply chain.

In addition there is now a degree of water recycling on the ISS. Nature's been recycling water on Earth for eons, and now NASA is set to do the same thing above Earth on the International Space Station. Space shuttle Endeavour carried in two refrigerator-sized racks packed with a distiller and an assortment of filters designed to process astronauts' urine and sweat into clean drinking water.

The station crew depends now on water carried up aboard a space shuttle or cargo rocket. But an operational water recycler is expected to cut that need by 65 percent by producing about 6,000 pounds of potable water each year. That's enough fresh water to allow the station to host six crew members instead of three.

A system that operates on the station also will provide a significant stepping stone to developing even more efficient processes that will support astronauts on the moon or on long-duration voyages into the solar system. Although Russia's space station Mir recycled cosmonaut's sweat, the NASA recycler is the first to be flown in space that intends to cleanse and reuse almost all the water a crew member produces.

The system can recycle about 93 percent of the water it receives, said Bob Bagdigian, the Environmental Control Life Support System project manager at NASA's Marshall Space Flight Center in Huntsville, Ala. The water recycler counts in large part on a distiller that Bagdigian compares to a keg tilted on its side. On Earth, distilling is a simple process of simply boiling water and cooling the steam back into pure water. But without gravity, the contaminants in water never separate from the steam no matter how much heat is used.

"In space, it becomes quite a challenge to distill any liquid in the absence of gravity," Bagdigian said.

So the keg-sized distiller is spun up to produce an artificial gravity field. The contaminants in the urine press against the sides of the drum while the steam gathers in the middle and is pumped to a filter. The filters are not much different from those used on Earth, which means they use charcoal-like materials to pull more unwanted elements from the water. Another process uses chemical compounds that bond with the remaining contaminants so filters can pick them out of the water, too.

"The water that we produce meets or exceeds most municipal water product standards," Bagdigian said. The system has been in different stages of development ever since NASA committed to building a space station in the 1980s. Along the way, individual parts of the system have been flown on space shuttle missions for tests.

The distiller mechanism flew in 2003 and worked just fine in orbit, Bagdigian said. Now the crew of the International Space

Station will test the whole apparatus, but they won't drink any at first. Instead, they will take numerous samples and return them to Earth for detailed testing. After the testing is complete, controllers will clear the astronauts to use the fresh water in orbit.

NASA's water filter development has also helped produce filters that are now used in humanitarian efforts to make clean water in areas served only by contaminated sources. The effort to make a crew support system that reduces the need for fresh supplies from Earth includes an oxygen generator that is already installed in NASA's Destiny lab on the space station.

Housed in one rack instead of the two required for the water recycler, the oxygen producer splits the oxygen and hydrogen molecules in water and sends the oxygen into the space station as breathable air. The hydrogen is now dumped overboard. However, another process is under development that will combine the hydrogen with other chemicals that react with each other and produce more water.

While the water recycler in use will work fine for the International Space Station's needs, Bagdigian said work is already under way to make it more efficient so it can be used on long moon exploration missions. "We'll take this system and continue to push its performance and efficiency," Bagdigian said.

Solar Array Wings

(The below systems can be adapted from ISS usage experiences)

The electrical system of the International Space Station is a critical resource for the ISS because it allows the crew to live comfortably, to safely operate the station, and to perform scientific experiments. The ISS electrical system uses solar cells to directly convert sunlight to electricity. Large numbers of cells are assembled in arrays to produce high power levels. This method of harnessing solar power is called photovoltaics.

The process of collecting sunlight, converting it to electricity, and managing and distributing this electricity builds up excess heat that can damage spacecraft equipment. This heat must be eliminated for reliable operation of the space station in orbit. The ISS power system uses radiators to dissipate the heat away from the spacecraft. The radiators are shaded from sunlight and aligned toward the cold void of deep space.

Each ISS solar array wing (often abbreviated "SAW") consists of two retractable "blankets" of solar cells with a mast between them. Each wing uses nearly 33,000 solar cells and when fully extended is 35 meters (115 ft.) in length and 12 meters (39 ft.) wide. When retracted, each wing folds into a solar array blanket box just 51 centimeters (20 in) high and 4.57 meters (15.0 ft.) in length. The ISS now has the full complement of eight solar array wings. Altogether, the arrays can generate 84 to 120 kilowatts.

The solar arrays normally track the Sun, with the "alpha gimbal" used as the primary rotation to follow the Sun as the space station moves around the Earth, and the "beta gimbal" used to adjust for the angle of the space station's orbit to the ecliptic. Several different tracking modes are used in operations, ranging from full Sun-tracking, to the drag-reduction mode ("Night glider" and "Sun slicer" modes), to a drag-maximization mode used to lower the altitude.

Batteries

Since the station is often not in direct sunlight, it relies on rechargeable nickel-hydrogen batteries to provide continuous power during the "eclipse" part of the orbit (35 minutes of every 90 minute orbit). The batteries ensure that the station is never without power to sustain life-support systems and experiments. During the sunlit part of the orbit, the batteries are recharged. The nickel-hydrogen batteries have a design life of 6.5 years which means that they must be replaced multiple times during the expected 20-year life of the station. The batteries and the battery charge/discharge units are manufactured by Space Systems/Loral (SS/L), under contract to Boeing. N-H2 batteries on the P6 truss were replaced in 2009 and 2010 with more N-H2 batteries brought by Space Shuttle missions. There are batteries in Trusses P6, S6, P4, and S4.

Since 2017, nickel-hydrogen batteries are being replaced by lithium-ion batteries. On January 6, a multi-hour EVA began the process of converting some of the oldest batteries on the ISS to the new lithium-ion batteries   There are a number of differences between the two battery technologies, and one difference is that the lithium-ion batteries can handle twice the charge, so only half as many lithium-ion batteries are needed during replacement. Also, the lithium-ion batteries are smaller than the older nickel-hydrogen batteries. Although they are not quite as long lasting as nickel-hydrogen, they can last long enough to extend the life of ISS.

ISS Electrical Power Distribution

The power management and distribution subsystem operates at a primary bus voltage set to Vmp, the peak power point of the solar arrays. As of 30 December 2005, Vmp was 160 volts DC (direct current). It can change over time as the arrays degrade from ionizing radiation. Microprocessor-controlled switches control the distribution of primary power throughout the station.

The battery charge/discharge units (BCDUs) regulate the amount of charge put into the battery. Each BCDU can regulate discharge current from two battery ORUs (Orbital Replacement Unit, a series-connected pack of 38 Ni-H2 cells), and can provide up to 6.6 kW to the Space Station. During insolation, the BCDU provides charge current to the batteries and controls the amount of battery overcharge. Each day, the BCDU and batteries undergo sixteen charge/discharge cycles. The Space Station has 24 BCDUs, each weighing 100 kg.

Data Architecture/Communications

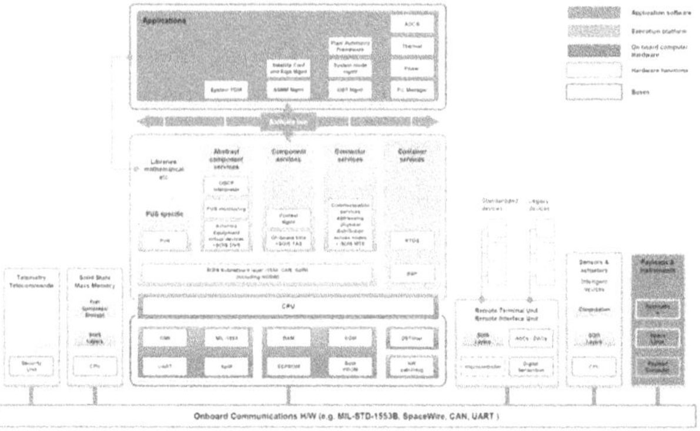

The ISS data architecture and communications system is very complex. I included an architecture diagram above and detailed overview below so that you can see just how much is involved.

(Similar systems can be used at the Moon Base.)

On a larger habitat in space imagine that the architecture is that much more complex according to its size and the number of people on it. Fortunately, computing architecture is one area where continuous advances should keep up with the computing needs of an advanced space habitat facility.

Spacecraft Management Unit

On the ISS The On-board Computer (also referred to as Spacecraft Management Unit - SMU or Command & Data Handling Management unit - CDMU) is the central core of the Spacecraft Avionics. The Central Processing Unit (CPU) hosts the

Execution Platform software (composed of RTOS, BSP, SOIS layers, PUS, …) and the Application software. Volatile and Non-volatile Memories, Safe Guard Memories, On Board Timer, Interface controllers and Reconfiguration modules are the other main blocks of a OBC. The figure above shows a functional architecture of the On-Board Data System where all the major functional blocks are indicated with their intercommunication links and their typical redundancy scheme.

Remote Terminal Unit

Remote Terminal Unit (also called Remote Interface Unit-RIU) is a unit that is usually present on medium-large size spacecraft. The RTU offloads the On Board Computer from analogue and discrete digital data acquisition and actuators control tasks.

Platform Solid State Mass Memory

For Earth Observation missions the mass memory for the P/L data may belong to the satellite platform and sometimes, depending on the capacity required, might be included inside the OBC as a single module.

TM/TC

The tele commands, once validated, are multiplexed to the intended addresses. There are two categories of commands: the high priority and the normal commands. The high priority commands (HPC) are sent to the Command Pulse Distribution Unit (CPDU) for immediate execution. The CPDU is either internal to the TC decoder or external and it's implemented in hardware, i.e. no software is involved in the execution of HPCs. The normal commands are sent off to the OBC CPU to be either processed or relayed on the system bus. The Telemetry encoder collects the Telemetry packets from different sources (processing, data storage, essential telemetry, payload), assembles the Telemetry transfer frames and sends them to the TM/TC transceiver to be downloaded to the ground.

Busses

The most common command and control bus used on a spacecraft platform is the MIL-STD-1553B covered by the ECSS-E-ST-50-13C. An alternative to the MIL-STD-1553B is the CAN that ESA and the European Space community is standardizing for space applications. UART serial channels are also used especially to control AOCS sensors. The Spacewire technology is now being increasingly used for data transfers < 160 Mbit/s and it can combine the command and control function with massive data transfer.

Communication protocols

The space community is asking for a real improvement in the specification and use of communications protocols. Typically, previous developments have harmonized physical interfaces and low level data link protocols but above this level proprietary solutions have been utilized. This has without any doubt increased development and integration costs and limited the possibility of element reuse without expensive modification. In comparison, the commercial market on the ground has systematically pursued the use of multilayer protocol stacks resulting in simple integration and multi-vendor compatibility. This commercial trend is now being adopted for the flight avionics by the development and standardization of protocols above the basic link layer.

Nuclear Power

Several government initiatives between NASA and Darpa are to produce new nuclear space propulsion systems which will also provide more potential for power production at moon bases.

19.0 Schedules for Construction

Schedules for a Lunar Base are still very preliminary and things will change over time. Here is one schedule document with speculations for future changes.

Since the below schedule was published the dates have been delayed a year or two due to the Covid pandemic.

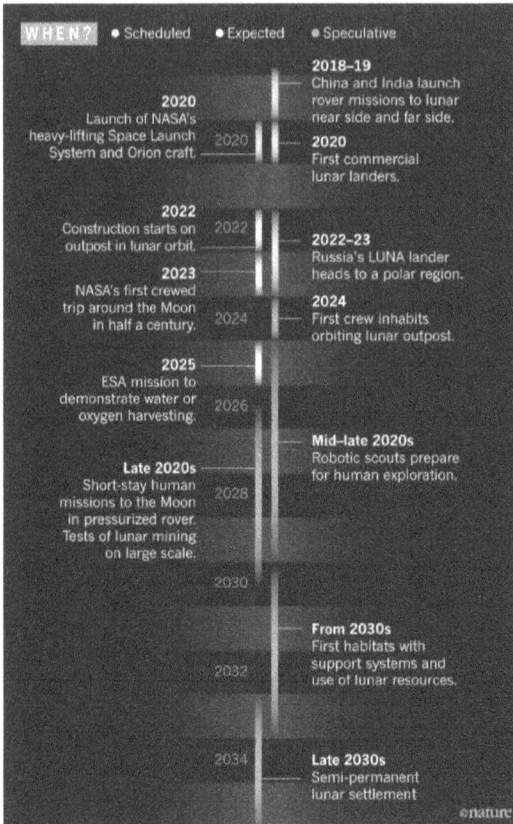

WHEN?   ● Scheduled   ● Expected   ● Speculative

2018–19
China and India launch rover missions to lunar near side and far side.

2020
Launch of NASA's heavy-lifting Space Launch System and Orion craft.

2020

2020
First commercial lunar landers.

2022
Construction starts on outpost in lunar orbit.

2022

2023
NASA's first crewed trip around the Moon in half a century.

2022–23
Russia's LUNA lander heads to a polar region.

2024

2024
First crew inhabits orbiting lunar outpost.

2025
ESA mission to demonstrate water or oxygen harvesting.

2026

Mid–late 2020s
Robotic scouts prepare for human exploration.

Late 2020s
Short-stay human missions to the Moon in pressurized rover. Tests of lunar mining on large scale.

2028

2030

From 2030s
First habitats with support systems and use of lunar resources.

2032

2034

Late 2030s
Semi-permanent lunar settlement

●nature

The first new Lunar Landing is scheduled for 2024 with a minimal orbiting sized Lunar Gateway to be used as a staging location for the lunar surface mission.

Successive visits to the Moon will then be scheduled yearly.

Lunar water harvesting will be tested in 2025

Short stays on the Moon will continue into the second half of the 2020s with first long term habitats to be built in the early 2030s.

20.0 Atreus Countries, Companies, and Plans

As of late 2022 over 25 countries have offered to participate in the Atreus Program as well as lots of major companies too. Here are some of the major initiatives of both member companies and countries:

20.1 Space Launch System

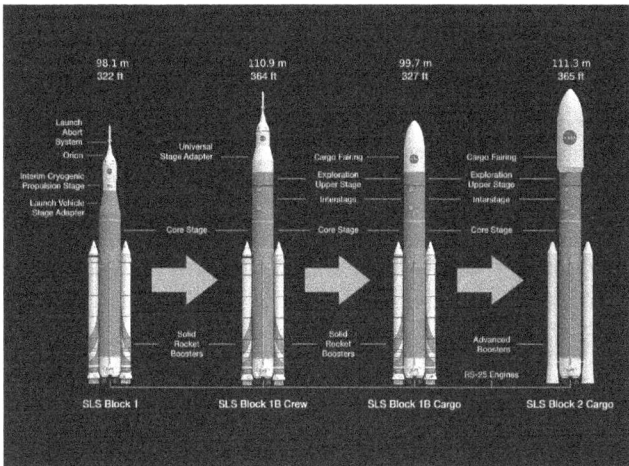

This system will launch the Lunar Gateway and Landers to the Moon

Boeing-They are the main contractor for the Space Launch System which is NASA's new heavy lift launch system to the Moon and beyond.

Space Launch System (SLS) Serious Cost Issues

The current design of the SLS is being rolled out for an unmanned test flight by mid 2022.

While the SLS seems to be a rocket that could provide the performance needed for man to visit the moon again, there are serious costs problems.

Current estimates in the Spring of 2022 is that each flight will cost over four billion dollars. And nothing on this rocketship is reusable.

On the other hand we look at SpaceX's Starship under development and it will cost around two million dollars per launch and is entirely reusable, and which will be rated to carry more cargo than the SLS.

That is an incredible cost difference of two thousand to one in terms of cost differences per flight. Once the starship starts flying, there is no way the continued usage of the SLS can be justified.

The SLS will become an obsolete white elephant very quickly since NASA will be able to use the money on other aspects of the Atreus missions in much more cost effective ways.

20.2 The Orion Capsule

This new deep space capsule can carry a crew of two to six persons and is designed for Moon and deep space missions.

Lockheed-Martin-They are the prime contractor for the Orion Capsule.

The Orion Multi-Purpose Crew Vehicle (Orion MPCV) is a class of partially reusable spacecraft used in NASA's human spaceflight programs. Consisting two components – a Crew Module (CM) manufactured by Lockheed Martin, and a European Service Module (ESM) manufactured by Airbus Defense and Space – the spacecraft are designed to support crewed exploration beyond low Earth orbit. Orion is equipped with solar power, an automated docking system, and glass cockpit interfaces modeled after those used in the Boeing 787 Dreamliner, and can support a crew of six up to 21 days undocked and up to six months docked. A single AJ10 engine provides the spacecraft's primary propulsion, while eight R-4 D-11 engines and six pods of custom reaction control system engines developed by Airbus provide the spacecraft's secondary propulsion. Although compatible with other launch vehicles, Orion is primarily designed to launch atop a Space Launch System (SLS) rocket, with a tower launch escape system.

Orion was conceived by Lockheed Martin as a proposal for the Crew Exploration Vehicle (CEV) to be used in NASA's Constellation program. Lockheed Martin's proposal defeated a competing proposal by Northrop Grumman, and was selected by NASA in 2006 to be the CEV. Originally designed with a service module featuring a new "Orion Main Engine" and a pair of circular solar panels, the spacecraft was to be launched atop the Ares I rocket with the Max Launch Abort System equipped. Following the cancellation of the Constellation program in 2010, Orion was heavily redesigned for use in NASA's Journey to Mars initiative; later named Moon to Mars.

The SLS replaced the Ares I as Orion's primary launch vehicle, and the service module was replaced with a design based on the European Space Agency's Automated Transfer Vehicle. A development test article of Orion's CM was launched in 2014 during Exploration Flight Test-1. As of 2019, two Orion spacecraft are under construction, with an additional two ordered, for use in NASA's Artemis program – the first of these is due to be launched in 2020 during Artemis 1.

ESA/Airbus-Defense-They will provide the service module for the Orion capsule which will have large wings to generate solar power and provide other services to Orion.

The European Service Module (ESM) is the service module component of the Orion spacecraft, serving as its primary power and propulsion component until it is discarded at the end of each mission. In January 2013, NASA announced that the European Space Agency (ESA) will contribute the service module for Artemis 1, based on ESA's Automated Transfer Vehicle (ATV). After approval of the first module, further decisions will be made that the module be provided by ESA for missions Artemis 1 to Artemis 4 included.

The service module supports the crew module from launch through separation prior to reentry. It provides in-space propulsion capability for orbital transfer, attitude control, and high altitude ascent aborts. It provides the water and oxygen needed for a habitable environment, generates and stores electrical power, and maintains the temperature of the vehicle's systems and components. This module can also transport unpressurized cargo and scientific payloads.

20.3 Lunar Gateway Partners

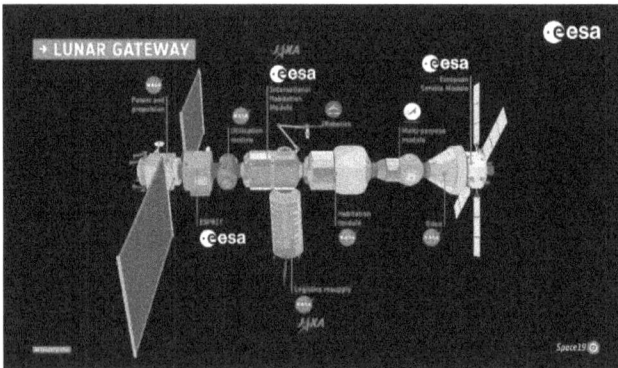

NASA-NASA will build a number of modules including:

The Power and Propulsion Element

The US Utilization Module

The US Habitation Module

Jaxa-The Japanese Space Agency will be helping on the International Habitation Module and the Logistics Resupply Module

Canada-They will provide a robotic arm like they already have for the Space Shuttle and the International Space Station

ESA-The European Space Agency is building several key items. These include the Service Module for the Orion Space Capsule, the

International Habitation Module for the Lunar Gateway, and the Esprit Module also for the Lunar Gateway.

Lunar Landers

There were multiple proposals for new lunar landers. SpaceX won the first round and NASA is looking at choosing a second lander too.

Blue Origin-Has offered one design for a Lunar Lander. This contract is still being bid and competed.

Lockheed-Martin-Has also proposed a Lunar Lander concept.

Additional Resource Providers

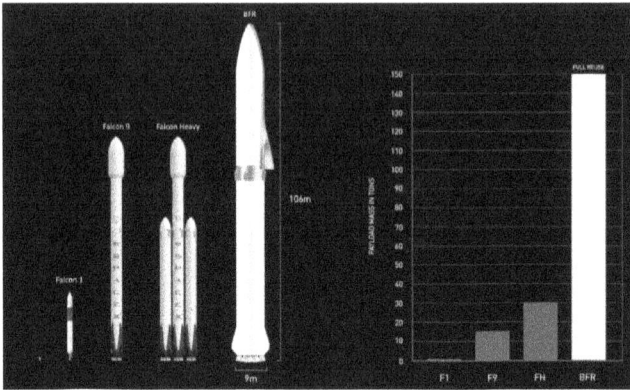

SpaceX-With its Falcon Heavy Rocket and planned BFR giant rocket plans to offer heavy lift capabilities for whatever additional transportation NASA needs for its Moon infrastructure and Moon Base.

24.0 Moon Base Cost Estimates

Estimated costs to ship materials to the Moon to start building the Moon Base will be very high. The Costs have been estimated in a very wide range.

Estimate A:

Some estimates done in 2007 say that reaching the Moon again would cost about $10 billion — estimates range from $7 billion and $13 billion — with an additional $28 billion to $52 billion being spent on the construction of base-related structures

Estimate B:

People have dreamt about living on the Moon for a long time, and while it is possible, it is also costly. Wendover Productions made a video explaining exactly how much it would cost to live on the Moon.

To calculate the amount, they figured out things such as how much a rocket would cost to use per pound, the cost of landers, the cost of a lunar base, and greenhouses for food. Their summary is that a grand total for four astronauts to live on the Moon for one year comes out to $36,000,000,000.

Estimate C:

In 2005, NASA estimated that returning humans to the Moon would cost $100 billion (approximately $122 billion in today's dollars). But if the success of private spaceflight companies like SpaceX and Orbital Sciences continues, NASA could send humans back to the lunar surface in as little as five to seven years, at a highly reduced cost, the new report shows.

That's not all: 10 to 12 years after that first commercial Moon trip, NASA could develop a permanent base on the Moon for about $40 billion in today's dollars, the report said. The proposed permanent Moon base would be used to convert lunar ice into hydrogen propellant that could be sold for use by other spacecraft, including missions headed to Mars.

Launch costs included in the report were based on prices quoted for SpaceX's Falcon 9 and Falcon Heavy rockets.

The command/service module for astronauts was based on the human-rated Dragon spacecraft that SpaceX is developing for International Space Station missions.

Cost Summary

The estimates are all over the map but we can expect that building a Moon Base would have costs similar to building the International Space Station which was $100 billion dollars.

25.0 A Story of Building the Moon Base

My book "The Moon and Beyond" is still a work in progress but it includes a few chapters with the most realistic description I could provide of the what the process of building a Moon Base would be like. Several sub chapters follow describing early Moon Base construction:

25.1 Early Moon Base Construction

We landed after several orbital corrections and a powered descent. The landing was anti-climactic with our engine kicking up lots of dust and then the motion stopped and we became aware that we were experiencing a one sixth Earth gravity. After making sure all the systems were operating properly, we opened the main hatch with us all wearing spacesuits to get outside.

The Commander and his assistant went down the ladder. Their first action was to activate the inflatable temporary structure at the base of the lander. This structure started to inflate and would provide us with shelter for the next several weeks as the main base was built. The inflatable dome was twenty feet in diameter and had an airlock built in. We each had a little sleeping cubicle and there was a galley and work areas. After it inflated we went inside to check airflow, heat, and then moved in supplies for living there.

We needed this shelter since living inside the lander for an extended period of time was a guarantee for crew stress and awful overcrowding.

We all took a walk around the landing site since we were all so excited to be there. The landing site was about one half mile from the crater wall and was in deep shade which was a couple of hundred degrees below zero Celsius. I could see the crater rim curving away to the horizon where it went out of sight. On the other side of the lander the land was just flat although we could see impact rocks in the distance. We could also see a glow over the crater rim because the Sun was shining on it from the other side.

I was also thinking about the best location for the shelter. Did we want it out in the open or next to the crater's wall? Next to the crater wall would be more protected in the long run. However, I didn't need to think long because the building's site had already been picked out on Earth.

That night we had a little party in the temporary structure and all rested well before work was to start in the morning.

Next morning Olga and I were the first ones outside. We needed to prepare the shelter site and setup the 3D construction equipment. There was a small tractor to be assembled which would be used to dig out a base for the building and make sure the foundation was firm. The tractor used a radio isotopic power source using Plutonium to create heat which was converted to electricity. The tractor had a radio control which we had practiced with back on Earth to control its movements.

I spent the next few hours digging a foundation pit with Olga relieving me as needed. After several days of effort we had a sufficient foundation dug and ready for construction. The foundation was round and one hundred feet in diameter. Our intent was to build a structure which could eventually hold fifty living and working people inside.

The next step was the construction of the larger girders for the three dimensional construction machine. Before we could actually start building the building we needed girders to raise the machine above the ground and provide tracks to move the construction printing head over the construction. Imagine that we were building a large printer larger than the building size. The printing head would move in computer controlled movements over the ground to print the building underneath it. The preparation project took several days.

We had been on the Moon a week and we were dead tired at the end of each workday. The commander made sure we all ate dinner together in the temporary shelter to update each other and build a sense of community.

Captains Hold and Neemar had just gotten back from a field trip up to the crater rim. Their job was to install a communications antenna with repeaters aimed to our site back inside the crater. They used an open rocket powered vehicle to launch up to the rim and come back. At the top in the sunlight, and in line with Earth they installed solar panels and the antenna and communications equipment we needed to have regular communications with the DSG and Earth.

Stark and Springer were responsible for the ice mining. They had already done radar surveys within a couple miles radius of our landing site using a simple lunar rover and were now laying out the foundation for the mining and extraction site. Another mission would bring sufficient mining and purification equipment to start the generation of large quantities of water, and its extraction into hydrogen and oxygen.

25.2 Main Shelter Construction

Finally, after almost two weeks of foundation digging, pre-construction, and setup of the 3D printing equipment we were ready to get started.

The machine had a hopper where we would feed it with Moon materials of a granular type, and a silicate based binder which would bind it all together like concrete.

We had the machine programmed to build a three level building with living quarters on the lowest (and safest) level, with office, labs, and manufacturing on the upper levels. There was even a garden area to grow vegetables to enhance our diet and produce some oxygen and filter out some carbon dioxide.

We turned on the machine and it started printing the bottom level of the building. All we had to do was keep feeding it the raw lunar regolith materials, the silicate binder, and electrical and piping which it would place and hold in locations as it poured walls and those materials became locked in place by the hardening walls. We wanted more solid materials like rebar to support the structure but didn't have the capacity to carry those materials to the Moon. Instead we were counting on the walls to be hard enough and carry enough load to make the structure solid.

Our building would be like those built by the Romans—who invented concrete. The Romans would use volcanic ash and lime in their concrete which they used to build many seaports and famous building like the Roman Colosseum. The Romans also didn't use rebar and most of their concrete structure were pure concrete like the Pantheon which was built with different thicknesses of concrete and well thought out geometric designs to give it the strength to hold up for two thousand years.

Over the next several days the building machine first printed a floor for the whole structure, then we could see the walls rising on the basement level which was designed to be twelve feet tall to give a sense of space. It would also have a hanging ceiling with air and other utilities in it which would make the visible height ten feet.

The plan was to finish and roof over the basement with its own airlock entrance as the rest of the building was completed. An elevator shaft was installed but blocked off temporarily. A ramp actually led up to the airlock on the first floor.

As the first level was completed Olga and I started moving in environmental systems and connecting them up. First was the air generation and ventilating system. It worked off of water ice which was now being produced from the ice mine. An automated supply ship also landed on Day 25. It homed in on a beacon our people had planted several hundred feet from our main base. The supply ship contained the rest of the initial ice mining and electrolysis equipment to produce usable quantities of fuel and other components we needed like air.

We also hooked up the waste recycling system. This system would dry out human waste and recycle the liquids. The dried waste could be used as fertilizer in the garden. It would be nice to use a toilet again rather than the tubes and bodily waste connections in the temporary inflatable shelter. We had to cut down the percentage of oxygen in the air to reduce fire risks and had brought tanks of liquid nitrogen which we installed with the air equipment to reduce the oxygen percentage to only twenty percent.

Each day Olga and I would take turns so that one of us was monitoring the construction machine while the other was working on systems in the building basement. After two more weeks the basement had a heater installed and air working inside. It also had a six inch thick ceiling to keep out solar radiation.

This was fortunate because the Commander called us all together early before dinner and we wondered what was up. He told us Earth was advising us of a large solar storm which would hit within six hours. It was projected to last several days and the inflatable tent would not provide enough protection. We would basically be barbequed if we stayed in the tent.

The main option was for us to live in the lander which was so compact nobody really wanted to do it. Both Olga and I suggested the basement of our structure was ready for occupation and would be safer than the lander because of its six inch thick ceiling. It would be ideal for a larger solar flare to have our full three foot thick roof. But given the projections of the flare, the current roof should work. Also due to the Moon's angle to the Sun and orbit around the Earth, we wouldn't be exposed to radiation problems for most of the storm. The Commander asked us more questions to be sure of the safety but we could see the relief on his face that we would not need to go live in the lander again for days.

Pretty soon he was assigning everyone tasks to move our food, sleeping equipment and more to the main shelter's basement. Over the next few hours we were a beehive of activity as we moved everything we could into the basement of the in progress building. That night we were all in the unfinished basement and it was pretty messy, but all of our life support systems were working properly.

We had radiation monitors all over the basement and the only area which registered dangerous was out next to the airlock upstairs.

We spent the next several days playing cards and watching movies while waiting for the solar storm to finish.

The next morning we were given the all clear by Earth and resumed construction. Now you could see the outline of the walls on the main level as the printed building continued to grow.

25.3 Finishing the Shelter

Olga and I restarted construction on the building that day. She worked outside while I worked on the interior of the basement level. This included setting up partitions for rooms. The pre-fab partitions were constructed outside by the construction machine. Then I would take them inside and position them for the rooms. They would then snap together to form walls and even doorways. The walls would fit in slots in the floor which were part of the original construction. I only had to do some drilling and screwing to connected power, doors, and more for each room.

As the rooms were built everyone started moving their personal items into them. Then I also started working on the kitchen and galley area. I got some help from other crew members who wanted this finished as quickly as possible.

Going outside that afternoon I could see that the walls were rising well on the main floor which contained some pre-defined rooms. The rest of the rooms would be based on movable partitions.

Away from the main shelter construction continued. The ice mining operation and splitting into component oxygen and hydrogen was now looking pretty close to completion.

Over the next few weeks the main floor of the shelter was finished and the smaller third floor was now under construction. You could see the main roof taking shape as the third floor grew. After another week of construction the third floor overall structure and roof was completed. The structure had no windows to keep the interior radiation safe. Windows would be simulated from exterior cameras which could display on large window screens inside.

My next big task was to run the tractor to push regolith over the roof. The idea was to bury it with at least several feet of covering to provide full protection well beyond any type of projected solar flare we could imagine.

I started by building a ramp on the side of the shelter away from the two airlocks. It took me most of that day to build the ramp. The next week was all about plowing regolith onto the roof. The regolith was then compacted in place by the building machine.

Finally, after over a month of construction you could look at the full outline of main shelter. It didn't look like much from the outside. All you could see from there was a big pile of soil with an entrance ramp and ramp cover going into a dark cavity. There were actually airlock ramps on two sides of the structure. In case one was needed as an emergency entrance.

When you entered the airlock you waited for the air pressure to be equalized. Then the door would open and you would enter into the equipment room with racks for spacesuits, and lockers with other outside equipment. Then a pressure tight door opened into the main shelter. At this location there were stairs up to the second level or down into the basement. A freight elevator was also next to the stairs to take larger equipment up and down. While the basement living area was pretty much finished, there was still a lot of construction on levels one and two. This construction would go on for months and the next crew rotation would also be continuing building.

26.0 Types of People Needed

The people needed to live and work at a Moon Base will have different backgrounds from our current Astronaut population. They will all probably be in their thirties and forties with previous experience in their careers.

Stays at the Moon Base will be similar to those at the Antarctic South Pole Base or International Space Station then get longer as capabilities of the base increase. One thing our new Moon visitors will probably not need to worry about is bone deterioration. One sixth gravity is light but it is a much healthier environment than microgravity. Still medical research on the long term effects of living on the Moon will need to be conducted.

Here are some of the potential types of work and careers for working and living on the Moon:

Pilots

NASA pilots will of course still be needed for the Orion Spacecraft, the Lunar Lander, and other equipment which needs guidance and control

Engineers

Engineers of all types will be needed to manage and repair high tech systems. I'm sure that there will also be many experiments which need a lot of in depth technical understanding to run.

Geologists

Understanding the Moon's different craters, Maria, and layering requires knowledge of how these things might be formed. Also to find the best locations to mine water ice at the South or North Pole.

Miners

Mining Water Ice is going to be very important for potential rocket fuel and water for the residents. Knowing mining techniques and how to use mining equipment will be very important.

Agronomists/Farmers

It's too expensive to transport all the food from the Earth to the Moon especially for long term residents. The ability to grow fruits and vegetables using hydroponics and lunar regolith for soil will help keep the costs down and offer a nice variety and fresh food for lunar residents.

Astronomers

With the Moon not having any atmosphere it will be a perfect location for lunar observatories. Astronomers would really enjoy the Moon for telescope locations.

Systems Engineers

There will be lots of data and communications networks at the Lunar Gateway and Moon Base and will need technical persons to setup, use, and maintain them. Lots of other high tech equipment will also be used and need in depth technical understanding to deploy, use, and maintain.

Construction Experience

The Moon Base will need initial and continuing building and enhancements. All the same specialty constructions skills used on Earth may also be needed on the Moon. Regular astronauts can perform some of the initial tasks which are well pre-defined, but eventually persons with the actual construction skills will be need for base maintenance and expansion.

Medical Doctors/Nurses

We will also need medical professionals to staff a Base since other medical treatment options on Earth will be days or even over a week away.

27.0 Reasons for a Moon Base

What are the reasons to build and support a Moon Base? In this chapter we provide some reasons:

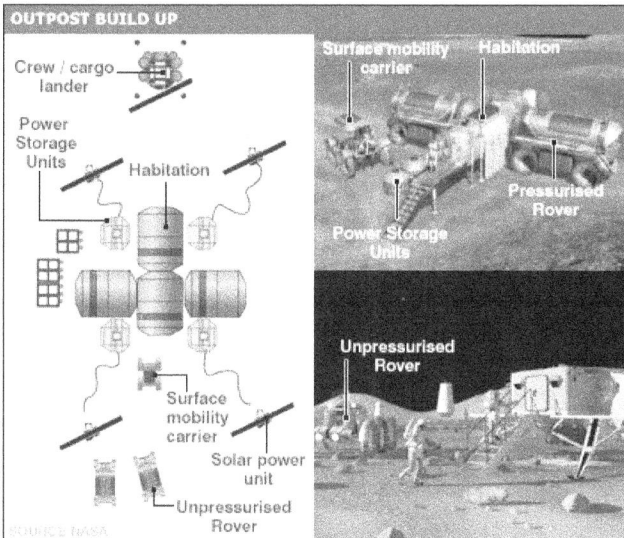

23.2 Scientific Research

A lunar base will create new opportunities for investigating the Moon and its environment and for using the Moon as a platform for scientific investigations. Analogous to the function of

McMurdo Base in Antarctica, the lunar base will provide logistical and supporting laboratory capability to rapidly expand knowledge of lunar geology, geophysics, environmental science, and resource potential through wide-ranging field investigations, sampling, and placement of instrumentation. Access to large, free vacuum volumes may enable new experimental facilities such as macro particle accelerators.

The fixed platform will enable new astronomical interferometric measurements to be obtained. The challenge of long-term, self-sufficient operations on the Moon can spur scientific and technological advances in materials science, bioprocessing, physics, and chemistry based on lunar materials, and reprocessing systems.

A lunar Farside observatory would also be feasible to build near a moon base at the poles. More information on Farside bases can be found in chapter 8.6.

27.2 A Dry run for Visiting Mars

NASA's stated purpose for the Atreus program is to place people on the Moon's surface and develop an ongoing presence there. This is all being done to put more infrastructure and plans in place for launching a manned mission to Mars.

So one of the main missions of a Moon base is to learn how to live, work, study, and generally learn how to exploit the resources of the Moon and apply that knowledge to a Mars exploration mission.

27.3 Exploitation of Lunar Resources

It has been argued that major industrialization of space cannot occur without access to the resources of the Moon. This might include immense projects such as solar power satellites.

A radio telescope located on the far side of the Moon would be shielded from background noise generated by terrestrial sources. An initial lunar instrument may well be a phased array of dipoles to be demonstrated at a sufficiently large scale. It is reasonable to develop the resource potential of the Moon to offset the high Earth-to-orbit transportation costs (Hearth, 1976). The lower gravitational field of the Moon and the absence of an atmosphere that retards objects accelerated from the surface provides a potential 20- to 30-fold advantage for launching from the Moon instead of Earth. For example, at liftoff, about 1.5%of the space shuttle's mass is payload. Most of the mass is propellant. From the Moon, approximately 50%of the mass can be payload.

The commodity currently envisioned to be most in demand in Earth-Moon space over the next three decades is liquid oxygen, which makes up 6/7 of the mass of propellant utilized by cryogenic (hydrogen-oxygen) rockets, such as the Centaur or postulated ones. Although it would appear unlikely that an atmosphere less body is a source for oxygen, it is actually an abundant element on the Moon (Arnold and Duke, 1978). It must be extracted, however, from silicate and oxide minerals into its liquid form for use as a propellant. Several processes have been suggested (Criswell, 1980) for accomplishing this, including reduction of raw soil by fluorine (which is recovered) or reduction of iron-titanium oxide (ilmenite) by hydrogen (also recovered). Preliminary laboratory studies have verified the concepts behind some of these processes.

Systems studies (e.g.,Carroll et al., 1983) show that oxygen production on the Moon could benefit STS in the early years of the next century, even if the hydrogen component of the propellant needed to be brought from Earth water at the lunar poles (Arnold, 1979) or extracting the dispersed solar wind-derived hydrogen in the lunar regolith would greatly improve the economics of the transportation system.

Other commodities also could be produced Metals, such as iron or titanium, can be extracted from the lunar soil or from specific rocks or minerals with differing degrees of difficulty. For example, small quantities of metal (primarily iron) from meteorites can be concentrated with a magnetic device from large amounts of lunar soil, or, with much larger energy inputs, titanium can be obtained from ilmenite. These products could find applications in large space structures. Lunar Titania or alumina might be used to produce aero brakes (heat shields). In the long term, at relatively high levels of development, production of components for solar electric power generation in space (e.g., solar power satellites) could be made feasible (Bock, 1979).

28.0 Moon Base Future Growth Plans

Once a Moon Base is established the plan is for the settlement to grow and expand to support research and other Moon related tasks.

Some plans have been made in the 1960s and 1970s for building a base and its plan for growth. Here are the possible stages from one report:

Phase I: Preparatory exploration

Lunar orbiter explorer and mapper
Instrument and experiment definition
Site selection
Automated site preparation

Phase II: Research outpost

Minimum base, temporarily occupied, totally resupplied from Earth
Small telescope/Geoscience module
Short range science sorties
Instrument package emplacement

Phase III: Operational base

Permanently occupied facility
Consumable production/Recycling pilot plant
Longer range science sorties
Geoscience and Medical laboratories
Experimental lunar radio telescope
Extended surface science experiment packages

Phase IV Advanced base

Advanced consumable production
Satellite outposts
Advanced geoscience laboratory
Plant research laboratory

Advanced astronomical observatory
Long-range surface exploration

Phase V: Self-sufficient colony

Full-scale production of exportable oxygen

Volatile production for agriculture, Moon-orbit transportation

Closed ecological life support system

Lunar manufacturing facility: tools, containment systems, fabricated assemblies, etc.

Lunar power station-100% lunar materials-derived

Expanding population base

29.0 Future Moon Structures

Below are some drawings for a future expansion of Moon Bases into a full city with a lot of it built underground.

Long term the Moon offers a good base for exploration of the Solar System. With its one sixth of an Earth gravity many things can land there for refueling and maintenance.

Also because of the lower gravity well, a linear accelerator could be built there to launch supplies and manned ships into orbit.

The Moon also offers options to give humanity another location to live in case of worldwide disasters on Earth.

30.0 Summary

Humanity will go back to the Moon and build a Moon Base. Current plans are now to have a return to the Moon landing set for 2025.

Lots of new technologies will be used to build the new Lunar Gateway and for the eventual Moon Base.

The technologies developed for the International Space Station will also help provide systems we can also use on the Moon Base.

Building a manned Lunar Base is a necessary step for a manned Mars mission and to send man to other objects in the Solar System.

I hope this book helps encourage many of you who are younger to participate in this exploration of our Solar System.

Martin K. Ettington

May 2022

31.0 Bibliography

1. Colonization of the Moon.
https://en.wikipedia.org/wiki/Colonization_of_the_Moon. [Online]

2. Moon EXploration Infographic.
https://www.space.com/26541-moon-exploration-350-year-history-infographic.html. [Online]

3. 2001 Movie Moon Base.
https://2001.fandom.com/wiki/Clavius_Base. [Online]

4. The Lunex Proposal.
https://en.wikipedia.org/wiki/Lunar_outpost_(NASA). [Online]

5. Project Horizon.
https://en.wikipedia.org/wiki/Project_Horizon. [Online]

6. Lunar Outpost.
https://en.wikipedia.org/wiki/Lunar_outpost_(NASA). [Online]

7. Shakelton Crater.
https://en.wikipedia.org/wiki/Shackleton_(crater). [Online]

8. Interesting Moon Base Proposal.
https://interestingengineering.com/8-interesting-moon-base-proposals-every-space-enthusiast-should-see. [Online]

9. 5 Fascinating Things You may not have known about the Moon. *https://www.techeblog.com/5-fascinating-things-that-you-may-not-have-known-about-the-moon/.* [Online]

10. Ambitious Plans to Colonize the Moon. *https://www.mentalfloss.com/article/53588/15-ambitious-plans-colonize-moon.* [Online]

11. Lunar Lava Tubes. *https://en.wikipedia.org/wiki/Lunar_lava_tube.* [Online]

12. Nasa secret plans for a Moon Base. *https://www.foxnews.com/science/secret-nasa-plans-for-moon-base-and-37-rocket-launches-revealed.* [Online]

13. *Lunar Base Concepts.* Institute, The Lunar and Planetary.

14. LESA Moon Base. *http://www.astronautix.com/l/lesalunarbase.html.* [Online]

15. NASA Lunder Lander Requirements. *https://spaceflightnow.com/2019/10/07/nasa-opens-competition-to-build-human-rated-lunar-landers/.* [Online]

16. How To Build a Moon Base. *https://www.nature.com/articles/d41586-018-07107-4.* [Online]

17. Cost to Live on the Moon. *http://www.astronomy.com/news/2016/09/how-much-it-would-cost-to-live-on-the-moon-in-9-minutes.* [Online]

18. Lunar Power Generation Study. *https://www.thespacereview.com/article/2882/1.* [Online]

19. Lunar Water Mining. *https://www.nasa.gov/directorates/spacetech/niac/2019_Phase_I_Phase_II/Lunar_Polar_Propellant_Mining_Outpost/.* [Online]

20. Lunar Exploration Partners. *https://spacenews.com/nasa-sees-strong-international-interest-in-lunar-exploration-plans/*. [Online]

21. https://www.space.com/russia-rekindle-moon-program-luna-25-launch. *Russia Rekindle Moon Program Luna 25 Launch.* [Online] 2022.

22. https://www.space.com/33440-space-law.html. *Space Law.* [Online] 2022.

23. https://www.smithsonianmag.com/air-space-magazine/next-robots-moon-180979366/#:~:text=NASA's%20plan%20to%20return%20to,generation%20of%20robotic%20lunar%20landers. *Robotic Lunar Landers.* [Online] 2022.

24. https://www.motortrend.com/news/gm-design-lockheed-martin-new-lunar-rover-renderings. *GM Design-Lockheed Martin New Lunar Rover Renderings.* [Online] 2022.

25. https://world-nuclear.org/information-library/non-power-nuclear-applications/transport/nuclear-reactors-for-space.aspx#:~:text=Radioisotope%20power%20sources%20have%20been,both%20the%20USA%20and%20Russia. *Nuclear Reactors for Space.* [Online] 2022.

26. https://www.theverge.com/2022/3/23/22993287/nasa-second-human-lunar-lander-moon-artemis-spacex. *NASA Second Humn Lunar Lander for Artemis moon program.* [Online] 2022.

27. https://www.newsweek.com/spacex-lunar-lander-nasa-seeks-more-ideas-artemis-moon-starship-1691301. *Spacex Luner Lander Design.* [Online] 2022.

28. https://www.space.com/china-russia-international-lunar-research-station. *China Russa International Research Station.* [Online] 2022.

29. *A_sustained_lunar_presence_nspc_report.* 2020.

30. https://spectrum.ieee.org/lunar-communications. *Lunar Communications.* [Online] 2022.

31. https://www.nasa.gov/mediacast/gravity-assist-how-to-grow-food-on-the-moon. *Gravity Assist-How to Grow Food on the Moon.* [Online]

32.0 Index

"Registration Convention, 21

(SLS) Serious Cost Issues, 168

2001 Movie, 34

Additional Resource Providers, 174

Agronomists/Farmers, 192

Andromeda satellite constellation, 108

Andy Weir, 36

ARTEMIS AFTER 2025, 139

Artemis Novel, 33

Artemis Program, 55

Arthur C. Clarke, 34

Astronomers, 192

Atmosphere, 12, 25

AVL is an industry leader, 102

Batteries, 150

Blue Origin, 174

Boeing, 168

Build, Fly, and Evolve, 59

Canada, 173

Chinese and Russian Lunar Base, 165

Chinese and Russian Lunar Plans, 163

Communication protocols, 163

Communications, 25

Contracted modules, 67

Cost Estimates, 176

Current Plans for Moon Bases, 153

Data Architecture/Communications, 161

Deep Space Gateway, 66

Dry run for Visiting Mars, 171

Earth and Lunar Environments, 11

Engineers, 191
ESA, 171
ESPRIT, 68
European System Providing Refueling, Infrastructure and Telecommunications, 68
Exploitation of Lunar Resources, 197
Extravehicular Pressure Suit Assembly, 88
Far Side Bases, 71
Far side of the Moon, 147
Finishing the Shelter, 188
First Moon Landings, 43
From the Earth to the Moon, 27
Future Growth, 199
Gateway Airlock Module, 69
Gateway Logistics Modules, 69
Geologists, 192
Gravity, 12
Growing Food, 26

Growing Food on the Moon, 124
H.G. Wells, 30
Habitation and Logistics Outpost, 68
HALO, 68
Heat and Cold, 169
iHAB, 69
Integrated Thermal Micrometeoroid Garment, 90
International Habitation Module, 69
ISS Electrical Power Distribution, 160
Jaxa, 173
Jules Verne, 27
Lava Tubes, 80
LESA Moon Base, 44
LGMO, 120
Liability Convention, 15
Liquid Cooling Garment, 91
Lockheed-Martin, 170, 174
Looking to the future, 229
Lunar Gateway Partners, 173

Lunar Ice, 15
Lunar Lander Concepts, 71
Lunar Orbit Details, 66
Lunar Polar Gas-Dynamic Mining Outpost, 120
Lunar Space Suits, 88
Lunar Transportation, 99
Lunar water harvesting, 166
Main Shelter Construction, 207
manmade tunnels, 82
Miners, 192
Mining Water Ice, 120
Moon Agreement, 21
Moon Base Construction, 179
NRHO orbit, 67
Nuclear Fission Power, 111
Orion Capsule, 170
Orion Crew Survival System, 94
Phase I: Preparatory exploration, 199

Phase II: Research outpost, 200
Phase III: Operational base, 200
Phase IV: Self-sufficient colony, 200
photovoltaic solar array, 104
Pilots, 191
Placing relay satellites, 110
Portable Life Support System, 91
Potential ice deposits, 17
Power and Propulsion Element, 67
Power Production, 104
Power Sources, 24
PPE, 67
Printing Buildings, 84
Project Horizon, 40
Radiation, 13
Radiation Protection, 24
Remote Terminal Unit, 162
Requirements, 24
Rescue Agreement, 21

Robert Heinlein, 32
Schedules, 165
Scientific Research, 194
Settling the South Pole, 78
Sewage, 26
Solar Array Wings, 158
Space Launch System, 168
Spacecraft Management Unit, 161
SpaceX, 175
SpaceX's Starship Lunar Lander, 75
Stanley Kubrick, 34
Subsurface moon bases, 42
Systems Engineers, 192
Temperature, 13
Temperature Control, 25
The First Man in the Moon, 30
The Moon is a Harsh Mistress, 32

The New Buggies, 101
The United Nations and the Outer Space Treaty, 19
TM/TC, 162
Tools and Equipment, 122
Torso Limb Suit Assembly, 89
Types of Moon Bases, 80
Types of People Needed, 191
Underground Base Plan, 47
Underground Structures, 86
Universal Space Transponder radio, 114
Unmanned Lander Plans, 52
Water, 13, 18
Water ice, 50
Working at the Moon Base, 219

Other books by Martin K. Ettington

Spiritual and
Metaphysics Books:

Prophecy: A History
and How to Guide

God Like Powers and
Abilities

Enlightenment for
Newbies

Removing Illusions to
Find True Happiness

Using the Scientific
Method to Study the
Paranormal

A Compendium of
Metaphysics and How to
Guides (Six books together
in one volume)

Love from the Heart

The Enlightenment
Experience

Learn Your Soul's
Purpose

Pursuing
Enlightenment

A Modern Man's Search
for Truth

Use Intuition and
Prophecy to Improve Your
Life

The Handbook of
Spiritual and Energy
Healing

Pure Spirituality and
God

Memories Before Birth
and Reincarnation

Paranormal Abilities
and the Yoga Sutras of
Patanjali

Mystical and Magical
Societies and Practitioners

Longevity &
Immortality:

Physical Immortality: A
History and How to Guide

The Commentaries of
Living Immortals

Records of Extremely
Long Lived Persons

Enlightenment and
Immortality

Longevity
Improvements from
Science

The 10 Principles of
Personal Longevity

Telomeres & Longevity

The Diets and Lifestyles
of the World's Oldest
Peoples

The Longevity Six
Books Bundle

Long Lived Plants and
Animals

A Guide to Longevity
Foods, Diets, and
Supplements

Science Fiction:
Out of This Universe

The Immortals of the
Interstellar Colony

The Mystic Soldier

The Immortality Sci Fi
Bundle

Visiting Many
Universes

The History of Science
Fiction and Fantasy

The God Like Powers
Series:

Human Invisibility

Invulnerability and
Shielding

Teleportation

Psychokinesis

Our Energy Body, Auras, and Thoughtforms

The God Like Powers Series—
Volume 1 Compilation

The Yoga Discovery Series:
Yoga-An Ancient Art Form
Hatha Yoga-Helping you Live Better
Raja Yoga-Through the Ages
The Yoga Discovery Package

Business & Coaching Books:
Creating, Paublishing, & Marketing Practitioner Ebooks
Building a Successful Longevity Coaching Business
Why Become a Coach?
The Professional Coaching Success Trilogy

2020-Make Money Writing and Selling Books
The 2020 Handbook of High Paying Work Without a College Degree
The important of Creativity and How to Improve Yours
Quantum Mechanics, Technology, Consciousness, and the Multiverse

Self-Improvement
Stress Relief and Methods to do so
The Importance of Creativity and How to Improve Yours
Building Self-Confidence
See the World Clearly
A Trilogy of Self Help Books
A New Paradigm of Truth and Happiness

Science, Technology, and Misc.

Future Predictions By and Engineer & Seer

The Unusual Science & Technology Bundle

Removing Limits On Our Consciousness-And Thinking Outside the Box

Universal Holistic Philosophy

Ball Lightning

Stranger Than Science Stories and Facts

Planet Earth is Conscious

Survival

Survival of Humanity Throughout the Ages

33 Incredible True Survival Stories

The Importance of Fire in History and Mythology

How to Survive Anything: From the Wilderness to Man Made Disasters

Building and Stocking a Nuclear Shelter for less than $10,000

The Human Survival Five Books Bundle

Stranger Than Science Facts and Stories

Stranger Than Science Facts and Stories Volume Two

Legendary Beings

Are Cryptozoological Animals Real or Imaginary?

Fire in History and Mythology

All About Dragons

Sea Serpents and Ocean Monsters

The Legendary Animals Five Books Bundle

The Mythical People of Ireland

Bigfoot Mysteries and Some Answers

About the Little People: Fairies, Elves, Dwarfs and Leprechauns

Ancient History

The Real Atlantis-In the Eye of the Sahara

Ancient & Prehistoric Civilizations

Ancient & Prehistoric Civilizations-Book Two

The History of Antediluvian Giants

The Antediluvian History of Earth

Ancient Underground Cities and Tunnels

Strange Objects Which Should Not Exist

More Out of Place Artifacts

Strange and Ancient Places in the USA

A Theory of Ancient Prehistory And Giant Aliens

The Destruction of Civilization About 10,500 B.C.

A Timeline of Intelligent Life on Earth

A 300 Million Year Old Civilization Existed on Earth

The Encyclopedia of Out of Place Artifacts

Aliens and Space

Types of UFOs Observed in History

Aliens and Secret Technology

Aliens Are Already Among Us

Designing and Building Space Colonies

Humanity and the Universe

All About Moon Bases

All About Mars Journeys and Settlement

The Space and Aliens Six Books Bundle

A Theory of Ancient Prehistory and Giant Aliens

The Space Colonies and Space Structures Coloring Book

All About Asteroids

Spaceships, Past, Present, and Future

Astronauts, Cosmonauts, and Other Important Space Flyers

All About Mars Journeys and Settlement

Mining the Asteroid Belt

The New Era of Space Stations

Time Travel and Dimensions

Real Time Travel Stories From a Psychic Engineer

The Real Nature of Time: An Analysis of Physics, Prophecy, and Time Travel Experiences

Stories of Parallel Dimensions

We Live in a Malleable Reality-and We Can Change It

The Time, Dimensions, and Quantum Mechanical Bundle

Alternate Dimensions & the Otherworld

Political and Social

The Empire of the United States: Forged By God's Spirit Through Man

The Longevity Training Series

(A transcription of the online Multimedia Longevity Coaching Training Program)

The Personal Longevity Training Series-Book1-Long Lived Persons

The Personal Longevity Training Series-Book2-Your Soul's Purpose

The Personal Longevity Training Series-Book3-Enable Your Life Urge

The Personal Longevity Training Series-Book4-Your Spiritual Connection

The Personal Longevity Training Series-Book5-Having Love in Your Heart

The Personal Longevity Training Series-Book6-Energy Body Health

The Personal Longevity Training Series-Book7-The Science of Longevity

The Personal Longevity Training Series-Book8-Physical Body Health

The Personal Longevity Training Series-Book9-Avoiding Accidents

The Personal Longevity Training Series-Book10-Implementing These Principles

The Personal Longevity Training Series-Books One Thru Ten

These books are all available in digital and printed formats from my

website and on Amazon, Barnes & Noble, Apple ITunes, and many other sites

My Books Website is: http://mkettingtonbooks.com

Signup for our Mailing List to get the following:

1) A discount coupon for 25% discount on all books on our site
2) Occasional Notices of new books available
3) Occasional Email on other offerings of ours (Monthly)

If you have any questions about this book or other subjects please contact the Author at:

mke@mkettingtonbooks.com